Kubernetes 零基础快速入门

张春晓 著

清華大學出版社
北京

内 容 简 介

Kubernetes 为容器化的应用提供了资源调度、部署、运行、服务发现、扩容和缩容等功能，充分发挥了容器技术的潜力，给开发人员和运维人员带来了极大的便利。本书由浅入深地讲解 Kubernetes 的功能，涵盖 Kubernetes 的核心组件，注重实战，能够帮助读者快速掌握在各种云平台上设计和部署大型集群的技能。

本书共 12 章，主要内容包括 Kubernetes 的安装方法、Kubernetes 命令行工具、应用系统的部署、通过服务访问应用、存储管理、软件包管理、网络管理、Dashboard（仪表盘）以及集群管理等，最后通过两个实战案例（部署 Spring Boot 应用、安装 KubeSphere）让读者上手实践 Kubernetes。

本书结构清晰、易教易学、实例丰富、可操作性强，对易混淆和实用性强的内容作了重点提示和讲解。本书适合使用 Kubernetes 的运维人员，也可以作为高等院校和培训机构计算机相关专业师生的教学参考书。

图书在版编目（CIP）数据

Kubernetes 零基础快速入门/张春晓著.—北京：清华大学出版社，2021.3（2024.2 重印）
ISBN 978-7-302-57625-9

Ⅰ. ①K… Ⅱ. ①张… Ⅲ. ①Linux 操作系统—程序设计 Ⅳ. ①TP316.85

中国版本图书馆 CIP 数据核字（2021）第 037420 号

责任编辑：夏毓彦
封面设计：王 翔
责任校对：闫秀华
责任印制：沈 露

出版发行：清华大学出版社
网　　址：https://www.tup.com.cn,https://www.wqxuetang.con
地　　址：北京清华大学学研大厦 A 座　　**邮　　编**：100084
社 总 机：010-83470000　　**邮　　购**：010-62786544
投稿与读者服务：010-62776969，c-service@tup.tsinghua.edu.cn
质量反馈：010-62772015，zhiliang@tup.tsinghua.edu.cn
印 装 者：三河市人民印务有限公司
经　　销：全国新华书店
开　　本：190mm×260mm　　**印　　张**：15.5　　**字　　数**：397 千字
版　　次：2021 年 4 月第 1 版　　**印　　次**：2024 年 2 月第 3 次印刷
定　　价：59.00 元

产品编号：090137-01

前　　言

云计算的概念从提出到今天，已经差不多 10 年了。在这 10 年间，云计算有了飞速的发展与翻天覆地的变化。如今，云计算被视为计算机网络技术领域的一次革命，因为它的出现，社会的工作方式和商业模式也在发生巨大的改变。

容器是云计算的核心，在一个大型的云中，会有大量的容器。据报道，Google 每周会启用超过 20 亿个容器。这么多容器的出现对人们容器管理的能力提出了巨大的挑战。

Kubernetes 的出现，为人们高效地管理和部署容器提供了一种便捷的手段。Kubernetes 可以提供所需的编排和管理功能，以便人们针对工作负载大规模部署容器。借助 Kubernetes 编排功能，用户可以构建跨多个容器的应用服务、跨集群调度、扩展这些容器，并长期持续管理这些容器的健康状况。有了 Kubernetes，用户便可以切实采取一些措施来提高 IT 安全性。

目前，在国外大部分的主流云服务提供商都采用了 Kubernetes 相关技术，包括亚马逊的 AWS、微软的 Azure。在国内，大部分的云服务提供商也提供了对于 Kubernetes 技术的支持，包括阿里云、腾讯云等。

然而，目前在国内人们对于 Kubernetes 的认识还远远不够，Kubernetes 的应用范围也相对比较小。

在 Kubernetes 发展过程中，出现了许多介绍容器和 Kubernetes 的书籍。但是，其中绝大部分都只是单独介绍上述几种技术，并没有将它们作为一个网站的整体解决方案来介绍。此外，大部分相关书籍的内容要么偏重理论，缺乏实践性；要么泛泛而论，缺乏深入的阐述。本书由具有丰富实践经验的一线技术人员编写而成，以实用为主旨，内容由浅入深，从最基本的入门知识开始，一直到比较深入的应用部署、存储管理、网络管理以及集群管理，系统地介绍了与 Kubernetes 整体方案相关的知识。

本书特色

1. 内容全面，系统性强

本书全面讨论了 Kubernetes 所涉及的各个方面，包括安装方法、核心概念、部署应用、服务管理、存储管理、网络管理以及集群负载均衡等。

2. 深入浅出，循序渐进

对于绝大部分的初学者来说，熟练运用 Kubernetes 是一件非常困难的事情。为了能够适应初学者的学习习惯，本书从最基本的基础知识开始讲起，一直到最后的网络管理、存储管理和集群管理，尽量从最简单的内容开始，逐步深入，避免初学者产生畏惧的心理。

3. 由一线技术人员编写，实践性、实用性强

本书内容的编写建立在大量实践经验的基础之上，因而具有很强的实用性。针对 Kubernetes 使用过程中最容易遇到的各种问题，依次展开论述。无论是初学者，还是具有一定经验的开发和维护人员，都可以从中获得有用的知识。

4. 重点突出，脉络清晰

对于比较重要的知识点，本书都进行非常深入的探讨，使得读者不仅知其然，更知其所以然，只有这样，才能使读者达到融会贯通的境界。

5. 项目案例典型，实战行强，有较高的应用价值

本书以目前最为流行的 Spring Boot 应用的部署以及 KubeSphere 安装部署为综合案例。这 2 个案例编码规范，使用广泛，具有很高的应用价值和参考价值。而且，这 2 个案例综合运用了本书所介绍的 Kubernetes 各方面的知识点，便于读者融会贯通地理解本书中所介绍的技术。此外，在介绍具体的技术过程中，本书也提供了大量具有实用参考价值的代码，这些代码稍加修改，便可用于实际项目开发中。

本书知识体系

第 1 章 Kubernetes 初步入门，介绍什么是 Kubernetes，以及 Kubernetes 的基本概念。

第 2 章安装 Kubernetes，介绍 Kubernetes 的安装方法，包括使用软件包管理工具进行安装、通过 kubeadmin 管理工具进行安装，以及通过二进制文件进行安装，甚至可以自行编译源代码再安装。

第 3 章 Kubernetes 命令行工具，介绍 Kubernetes 提供的主要命令行工具，主要包括 kubeadm 和 kubectl 等。

第 4 章运行应用，详细介绍如何在 Kubernetes 中部署各种容器化应用。

第 5 章通过服务访问应用，介绍服务的管理方法以及如何通过 ClusterIP、NodePort 和 LoadBalancer 这三种方式来访问应用系统。

第 6 章存储管理，详细介绍 Kubernetes 的存储系统。

第 7 章 Kubernetes 软件包管理，介绍 Helm 的使用方法。

第 8 章 Kubernetes 网络管理，详细介绍 Kubernetes 的网络基础知识以及各种网络方案，并重点介绍 Flannel 的使用方法。

第 9 章 Kubernetes Dashboard，详细介绍 Kubernetes Dashboard 的安装方法以及如何通过 Dashboard 来管理集群。

第 10 章 Kubernetes 集群管理，详细介绍 Kubernetes 集群中各种资源的管理方法以及 Pod 的驱逐机制和高可用部署等。

第 11 章在 Kubernetes 集群中部署 Spring Boot 应用程序实战，以一个具体的应用系统为例，介绍如何在 Kubernetes 集群中部署 Spring Boot 应用系统。

第 12 章安装 KubeSphere 实战，详细介绍 KubeSphere 的安装和使用方法。

适合阅读本书的读者

- 需要全面学习 Kubernetes 系统维护、开发及云计算技术的人员
- 网络架构工程师
- 云计算咨询顾问
- IT 管理人员
- 高校和专业培训机构的师生
- 需要一本案头必备查询手册的人员

示例源码下载

本书配套的示例源代码下载，请用微信扫描右边二维码获取，可按扫描结果页面的提示，转发到自己的邮箱中下载。如果有任何问题，请直接发邮件至 booksaga@163.com，邮件主题为“Kubernetes 零基础快速入门”。

作　者

2021 年 1 月

目　　录

第 1 章

Kubernetes初步入门

为了使读者快速了解 Kubernetes，本章将介绍什么是 Kubernetes，以及 Kubernetes 的基本概念。

本章涉及的知识点有：

- Kubernetes：主要介绍 Kubernetes 的基本概念，Kubernetes 的发展历史以及为什么使用 Kubernetes 等。
- Kubernetes 的重要概念：主要包括 Cluster、Master、Pod、卷、服务以及命名空间等。

1.1 Kubernetes 技术

Kubernetes 是容器技术快速发展的产物。Kubernetes 的出现使得大规模的服务器运维变得便捷、简单起来。本节先从最简单的概念开始，向读者顺序渐进地介绍 Kubernetes。

1.1.1 什么是 Kubernetes

简单地讲，Kubernetes 是一套自动化容器运维的开源平台，这些运维操作包括部署、调度和节点集群间扩展。对比 Docker 技术来看，可以将 Docker 看作是 Kubernetes 内部使用的低级别的组件，而 Kubernetes 则是管理 Docker 容器的工具。如果把 Docker 容器比作是飞机，则 Kubernetes 可比作是飞机场。

1.1.2 Kubernetes 的发展历史

早在十年前，Google 就开始大规模地使用容器技术。据说，Google 的数据中心运行着 20 多亿个容器。管理好如此数量庞大的容器是一件非常困难的事情。因此，Google 开发了一套叫作 Borg 的系统来对容器进行调度和管理。

在经过多年的实践、经验积累和改进之后，Google 重新编写了这套容器管理系统，并且将其贡献给开源社区，这就是 Kubernetes 的来源。Kubernetes 一经开源就一鸣惊人，并迅速在容器技术领域称霸。

在刚开源的前两年，Kubernetes 只有五个主要版本。从 2017 年起，它相继推出了 1.6、1.7、1.8、1.9，围绕稳定性以及性能做了改进。而在 2018 年，Kubernetes 更进一步，又进行了 4 次重大更新，在企业最关注的安全性和可扩展性上做了显著改善。

2018 年 3 月 27 日，Kubernetes v1.10 发布。此版本持续增强了 Kubernetes 的成熟性、可扩展性以及可插拔性，并在存储、安全、网络增强了其稳定性。

2018 年 6 月 28 日，Kubernetes v1.11 发布。此版本增强了网络功能、可扩展性与灵活性。Kubernetes v1.11 功能的更新为任何基础架构、云或内部部署都能嵌入到 Kubernetes 系统中增添了更多可能性。

2018 年 9 月 28 日，Kubernetes v1.12 发布。此版本新增了两个备受期待的功能，Kubelet TLS Bootstrap 和对 Azure 虚拟机规模集支持（并已达到 GA 阶段）。同时，该版本在安全性和 Azure 等关键功能上做出了改进。

2018 年 12 月 4 日，Kubernetes v1.13 发布。此版本是迄今为止发布时间最短的版本之一。此版本中的显著特征包括：kubeadm 简化集群管理、Container Storage Interface（CSI）、CoreDNS 为默认 DNS。这三个主要特性在这个版本中已逐渐过渡到 GA。

截至 2020 年 12 月，Kubernetes 发布了 v1.20.0。

Kubernetes 所处的时间点是传统和现代软件开发日益高流量的交叉点。根据 CNCF 统计数据，2018 年，云原生技术增长了 200%，全球有近三分之一的企业正在运营多达 50 个容器，运营 50~249 个容器的企业占比也超过 25%，有超过 80%的受访者把 Kubernetes 作为容器管理的首选。

Kubernetes 作为当前唯一被业界广泛认可和看好的 Docker 分布式系统解决方案，可以预见，在未来几年内，会有大量的新系统选择它。容器化技术已经成为计算模型演化的一个开端，Kubernetes 作为容器开端的 Docker 容器集群管理技术，在这场新的技术革命中扮演着重要的角色，并具有不可预估的发展前景和商业价值。

1.1.3 为什么使用 Kubernetes

Kubernetes 是一个自动化部署、伸缩和操作应用程序容器的开源平台。使用 Kubernetes 可以快速、高效地满足用户的以下需求：

- 快速精准地部署应用程序。
- 即时伸缩你的应用程序。
- 无缝展现新特征。
- 限制硬件用量仅为所需资源。

Kubernetes 具有以下明显的优势：

- 可移动：公有云、私有云、混合云、多态云。
- 可扩展：模块化、插件化、可挂载、可组合。
- 自修复：自动部署、自动重启、自动复制、自动伸缩。

为什么我们需要 Kubernetes，它能做什么？

至少，Kubernetes 能在实体机或虚拟机集群上调度和运行程序容器。而且，Kubernetes 也能让开发者斩断联系着实体机或虚拟机的“锁链”，从以主机为中心的架构，跃至以容器为中心的架构。该架构最终提供给开发者诸多内在的优势和便利。Kubernetes 提供给基础架构以真正的以容器为中心的开发环境。

Kubernetes 满足了一系列产品内运行程序的共同需求，诸如：

- 协调辅助进程，协助应用程序整合，维护一对一“程序 - 镜像”模型。
- 挂载存储系统。
- 分布式机密信息。
- 检查程序状态。
- 复制应用实例。
- 负载均衡。
- 滚动更新。
- 资源监控。
- 访问并读取日志。
- 程序调试。
- 提供验证与授权。

1.2　Kubernetes 重要概念

了解和掌握 Kubernetes 中的重要概念是非常重要的。只有深入理解 Kubernetes 的基本概念，才能掌握各个组件的功能，从而能够熟练地部署和维护 Kubernetes。本节将对 Kubernetes 中的重要概念进行介绍。

1.2.1　Cluster

在 Kubernetes 中，Cluster 是计算、存储和网络资源的集合。Kubernetes 利用这些基础资源来运行各种应用程序。因此，Cluster 是整个 Kubernetes 容器集群的基础环境。

1.2.2　Master

Master 是指集群的控制节点。在每个 Kubernetes 集群中，都至少有一个 Master 节点来负责整个集群的管理和控制。几乎所有的集群控制命令，都是在 Master 上面执行的。因此，Master 是整个集群的大脑。正因为 Master 如此重要，所以为了实现高可用性，用户可以部署多个 Master 节点。Master 节点可以是物理机，也可以是虚拟机。

关于 Master 上运行的关键进程，下面进行逐个说明。

1. Kubernetes API Server

Kubernetes API Server 的进程名为 kube-apiserver。Kubernetes API Server 提供了 Kubernetes 各类资源对象的增、删、改、查的 HTTP Rest 接口，是整个系统的数据总线和数据中心。Kubernetes API Server 提供了集群管理的 REST API 接口，包括认证授权、数据校验以及集群状态变更，提供了其他模块之间的数据交互和通信的枢纽，是资源配额控制的入口，拥有完备的集群安全机制。

2. Kubernetes Controller Manager

Controller Manager 作为集群内部的管理控制中心，负责集群内的 Node、Pod 副本、服务端点（Endpoint）、命名空间（Namespace）、服务账号（ServiceAccount）、资源配额（ResourceQuota）的管理。当某个 Node 意外宕机时，Controller Manager 会及时发现，并执行自动化修复流程，确保集群始终处于预期的工作状态。

3. Kubernetes Scheduler

Kubernetes Scheduler 的作用是根据特定的调度算法 Pod 调度到指定的工作节点（Node）上，这一过程也叫绑定（Bind）。Scheduler 的输入为需要调度的 Pod 和可以被调度的节点（Node）的信息，输出为调度算法选择的 Node，并将该 Pod 绑定到这个 Node。

4. Etcd

Etcd 是 Kubernetes 集群中的一个十分重要的组件，用于保存集群所有的网络配置和对象的状态信息。

1.2.3 Node

在 Kubernetes 中，除了 Master 节点之外，其他的节点都称为 Node。与 Master 节点不同，Node 才是 Kubernetes 中的承担主要计算功能的工作节点。Node 可以是一台物理机，也可以是一台虚拟机。

整个 Kubernetes 集群中的 Node 协同工作，Master 会根据实际情况将某些负载分配给各个 Node。当某个 Node 出现故障时，其他的 Node 会替代其功能。

Node 节点将运行以下主要进程：

1. Kubelet

在 Kubernetes 集群中，每个 Node 节点都会启动 kubelet 进程，用来处理 Master 节点下发到本节点的任务，管理 Pod 和其中的容器。Kubelet 会在 API Server 上注册节点信息，定期向 Master 汇报节点资源使用情况，并通过 cAdvisor 监控容器和节点资源。可以把 Kubelet 理解成是一个代理进程，是 Node 上的 Pod 管家。

2. kube-proxy

kube-proxy 运行在所有 Node 节点上，它监听每个节点上 Kubernetes API 中定义的服务变化情况，创建路由规则来进行服务负载均衡。

3. Docker 引擎

该 Docker 引擎指 Docker CE 服务引擎，负责容器的创建和管理等工作。

1.2.4　Pod

Pod 是 Kubernetes 最基本的操作单元，一个 Pod 中可以包含一个或多个紧密相关的容器，一个 Pod 可以被一个容器化的环境看作应用层的逻辑宿主机。一个 Pod 中的多个容器应用通常是紧密耦合的，Pod 在 Node 上被创建、启动或者销毁。每个 Pod 里运行着一个特殊的被称之为 Pause 的容器，其他容器则为业务容器，这些业务容器共享 Pause 容器的网络栈和 Volume 挂载卷，因此它们之间通信和数据交换更为高效。在设计时我们可以充分利用这一特性，将一组密切相关的服务进程放入同一个 Pod 中。

同一个 Pod 里的容器之间仅需通过 localhost 就能互相通信。同一个 Pod 里面的业务容器共享 Pause 容器的 IP 地址，共享 Pause 容器挂载的存储卷。

Pod 是 Kubernetes 调度的基本工作单元，Master 节点会以 Pod 为单位，将其调度到 Node 节点上面。Pod 的基本组成如图 1-1 所示。

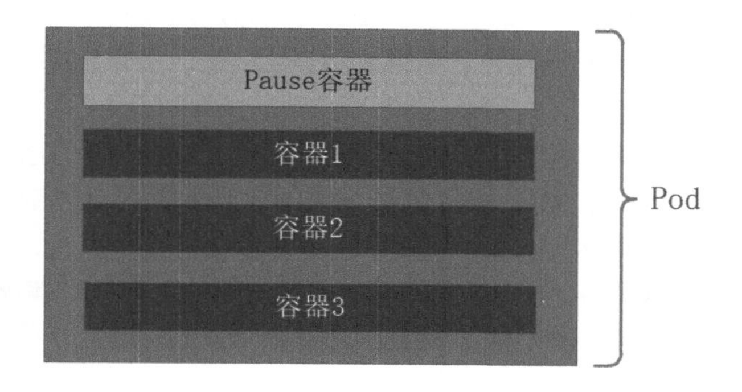

图 1-1　Pod 的基本组成

1.2.5　服务

在 Kubernetes 的集群中，虽然每个 Pod 都会被分配一个单独的 IP 地址，但这个 IP 地址会随着 Pod 的销毁而消失。这就引出一个问题，如果有一组 Pod 组成一个集群来提供服务，那么如何来访问它呢？那就是通过服务。

一个服务可以看作一组提供相同服务的 Pod 的对外访问接口，服务作用于哪些 Pod 是通过标签选择器来定义的。服务通常拥有以下特点：

- 拥有一个指定的名字，比如 mysql-server。
- 拥有一个虚拟 IP 地址和端口号，销毁之前不会改变，只能内网访问。
- 能够提供某种远程服务能力。
- 被映射到了提供这种服务能力的一组容器应用上。

如果服务要提供外网服务，则需指定公共 IP 和 Node 端口，或外部负载均衡器。

1.2.6 卷

默认情况下，容器的数据都是非持久化的，在容器消亡以后数据也跟着丢失，所以 Docker 提供了卷机制以便将数据持久化存储。类似地，Kubernetes 提供了更强大的卷机制和丰富的插件，解决了容器数据持久化和容器间共享数据的问题。

与 Docker 不同，Kubernetes 卷的生命周期与 Pod 绑定。容器宕掉后 Kubelet 再次重启容器时，卷的数据依然还在，而 Pod 删除时，卷才会清理。数据是否丢失取决于具体的卷类型，比如 emptyDir 类型的卷实际上是一个临时空目录，为 Pod 内多用户同享一个目录。与 Pod 的生命周期一致，Pod 创建时创建，删除时删除。持久化存储卷为独立于计算资源的一种实体存储资源，不属于任何一个 Node。因此，在 Pod 被删除时，不会丢失数据，除非人工将其删除。

1.2.7 命名空间

命名空间是 Kubernetes 系统中的另一个重要的概念，通过将系统内部的对象分配到不同的命名空间中，形成逻辑上的不同项目、小组或用户组，从而使得在共享使用整个集群的资源的同时还能被分别管理。

Kubernetes 集群在启动后，会创建一个名为 default 的默认的命名空间，如果不特别指明命名空间，则用户创建的 Pod、RC、服务都被系统创建到默认的命名空间中。

当团队或项目中具有许多用户时，可以考虑使用命名空间来区分。在未来的 Kubernetes 版本中，默认情况下，相同命名空间中的对象将具有相同的访问控制策略。

第 2 章

安装Kubernetes

在第 1 章中，读者已经对 Kubernetes 有了初步的了解。接下来，本章将介绍 Kubernetes 的安装方法。Kubernetes 的安装方式非常灵活，用户可以通过软件包管理工具进行安装，也可以通过 kubeadmin 管理工具进行安装，或者通过二进制文件进行安装，甚至可以通过源代码自己编译安装。本章将对 Kubernetes 的主要安装方式进行介绍。

本章涉及的知识点有：

- 通过软件包管理工具安装 Kubernetes：主要介绍如何通过 CentOS 的 yum 软件包管理工具安装 Kubernetes。
- 通过二进制文件安装 Kubernetes：介绍如何通过 Kubernetes 官方提供的二进制文件进行 Kubernetes 的安装部署。
- 通过源代码安装 Kubernetes：介绍如何获取 Kubernetes 源代码，以及如何进行编译和安装部署。

2.1 通过软件包管理工具安装 Kubernetes

Kubernetes 为绝大部分的操作系统平台都提供了相应的软件包。通过软件包来安装 Kubernetes 是一种最简单的安装方式。对于初学者来说，通过这种方式可以快速搭建起 Kubernetes 的运行环境。本节将以 CentOS 为例，介绍如何通过软件包来安装 Kubernetes。

2.1.1 软件包管理工具

大多数现代的 Linux 发行版都提供了一种中心化的机制，用来搜索和安装软件。软件通常都是存放在中心存储库中，并通过包的形式进行分发。处理包的工作被称为包管理。包提供了操作系统的基本组件，以及共享的库、应用程序、服务和文档。

目前，在 Linux 系统下常见的软件包格式主要有：RPM 包、TAR 包、bz2 包、gz 包以及 deb 包等。其中 RPM 包是最常见的 Linux 软件包形式，由美国 RedHat 公司开发，最初在其发布的 RedHat Linux 发行版中使用。目前 RPM 已经是 Linux 的软件包管理标准。TAR 包是 Linux 的一种文件归档形式，非常多的文件都以 TAR 包形式压缩打包。bz2 和 gz 都是比较常见的压缩包格式。deb 则是 Debian 的制订的软件包管理形式。

在 CentOS 中，用户可以通过 rpm 命令或者 yum 命令来安装软件包，而 yum 则是最常用的软件包管理工具。yum 命令的基本语法如下：

```
yum [options] command
```

yum 命令的选项比较多，其中最常用的主要有两个，其中一个是-skip-broken，该选项的功能是在安装指定软件包的时候，忽略依赖检查。尽管在某些特殊情况下，使用该选项可以将软件包安装上去，但是往往会导致软件包不能正常工作。因此，用户应该尽量避免使用该选项。另外一个选项为-y，由于在默认情况下，yum 采用交互式安装软件包，因此，在安装过程中会不断询问用户一些问题，要求用户回答 yes 或者 no。如果遇到问题比较多，的确会令人觉得非常烦琐。此时，用户可以使用-y 选项，这意味着对于所有需要回答 yes 或者 no 的问题，一律自动选择 yes。

常用的命令有 erase、install 以及 search 等。其中，erase 命令用来将软件包从当前系统中删除，install 命令用来安装指定的软件包，search 命令用来搜索指定的软件包。这 3 个命令的语法相同，后面加上软件包名称即可。

2.1.2 节点规划

在本例中，部署了 3 台主机，其中 1 台为 Master 节点，另外 2 台为 Node 节点，其网络拓扑如图 2-1 所示。

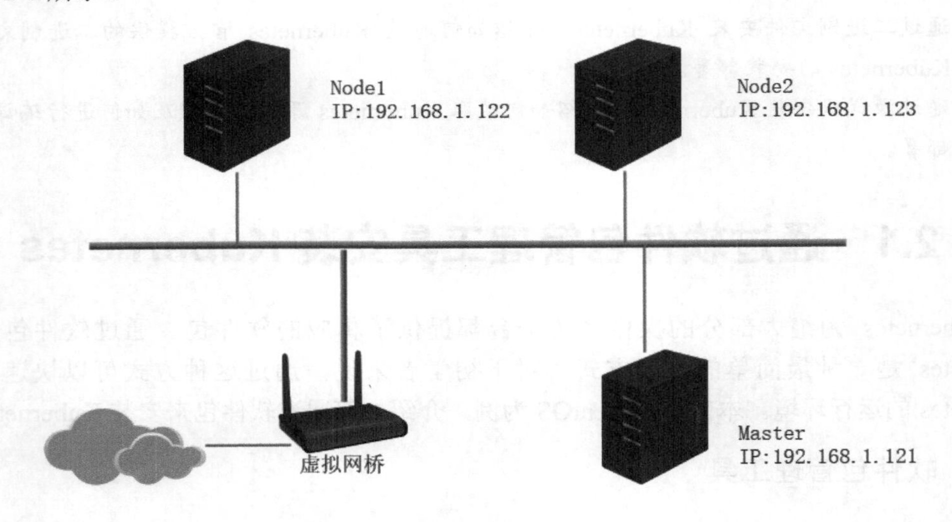

图 2-1 网络拓扑图

其中 Master 节点上面安装 kubernetes-master 和 etcd 软件包，Node 节点上面安装 kubernetes-node、etcd、Flannel 以及 Docker 等软件包。kubernetes-master 软件包包括 kube-apiserver、kube-controller-manager 以及 kube-scheduler 等组件及其管理工具。kubernetes-node 包括 kubelet 以及管理工具。2 个 Node 节点上面都安装 etcd 软件包，这样 3 个节点组成一个 etcd 集群。Flannel 为网络组件，Docker 运行容器。

2.1.3　安装前准备

在安装软件包之前，首先对所有节点的软件环境进行相应的配置和更新。

1. 禁用 SELinux

SELinux 是 2.6 版本的 Linux 内核中提供的强制访问控制系统，在这种访问控制体系的限制下，进程只能访问某些指定的文件。尽管 SELinux 可以在很大程度上加强 Linux 的安全性，但是它会影响 Kubernetes 某些组件的功能，所以需要将其禁用，命令如下：

```
[root@localhost ~]# setenforce 0
```

以上命令仅仅暂时禁用了 SELinux，在系统重启之后，SELinux 又会重新发挥作用。为了彻底禁用 SELinux，用户需要修改其配置文件/etc/selinux/config。将其中的

```
SELINUX=enforcing
```

修改为

```
SELINUX=disabled
```

2. 禁用 firewalld

firewalld 是 CentOS 7 开始采用的防火墙系统，代替之前的 iptables。用户可以通过 firewalld 来加强系统的安全，关闭或者开放某些端口。firewalld 会影响 Docker 的网络功能，所以在安装部署前需要将其禁用，命令如下：

```
[root@localhost ~]# systemctl stop firewalld
[root@localhost ~]# systemctl disable firewalld
```

3. 更新软件包

在安装部署 Kubernetes 之前，用户应该更新当前系统的软件包，保持所有的软件包都是最新版本，命令如下：

```
[root@localhost ~]# yum -y update
```

4. 同步系统时间

用户需要先将三台服务器的时间通过 NTP 进行同步，否则，在后面的运行中可能会提示错误，命令如下：

```
[root@localhost ~]# ntpdate -u cn.pool.ntp.org
```

其中 cn.pool.ntp.org 为中国的网络时间协议（NTP）服务器。

2.1.4　etcd 集群配置

前面已经介绍过，etcd 是一个高可用的分布式键值数据库。Kubernetes 利用 etcd 来存储某

些数据。为了提高可用性，在本例中我们在 3 台服务器上面都部署 etcd，形成一个拥有 3 个节点的集群。

在 Master 节点上面执行以下命令：

```
[root@localhost ~]# yum -y install kubernetes-master etcd
```

然后修改 etcd 配置文件/etc/etcd/etcd.conf，内容如下：

```
#[Member]
#ETCD_CORS=""
ETCD_DATA_DIR="/var/lib/etcd/default.etcd"
#ETCD_WAL_DIR=""
ETCD_LISTEN_PEER_URLS="http://192.168.1.121:2380"
ETCD_LISTEN_CLIENT_URLS="http://192.168.1.121:2379,http://127.0.0.1:2379"
#ETCD_MAX_SNAPSHOTS="5"
#ETCD_MAX_WALS="5"
ETCD_NAME="etcd1"
#ETCD_SNAPSHOT_COUNT="100000"
#ETCD_HEARTBEAT_INTERVAL="100"
#ETCD_ELECTION_TIMEOUT="1000"
#ETCD_QUOTA_BACKEND_BYTES="0"
#ETCD_MAX_REQUEST_BYTES="1572864"
#ETCD_GRPC_KEEPALIVE_MIN_TIME="5s"
#ETCD_GRPC_KEEPALIVE_INTERVAL="2h0m0s"
#ETCD_GRPC_KEEPALIVE_TIMEOUT="20s"
#
#[Clustering]
ETCD_INITIAL_ADVERTISE_PEER_URLS="http://192.168.1.121:2380"
ETCD_ADVERTISE_CLIENT_URLS="http://192.168.1.121:2379"
#ETCD_DISCOVERY=""
#ETCD_DISCOVERY_FALLBACK="proxy"
#ETCD_DISCOVERY_PROXY=""
#ETCD_DISCOVERY_SRV=""
ETCD_INITIAL_CLUSTER="etcd1=http://192.168.1.121:2380,http://192.168.1.12
2:2380,http://192.168.1.123:2380"
#ETCD_INITIAL_CLUSTER_TOKEN="etcd-cluster"
#ETCD_INITIAL_CLUSTER_STATE="new"
#ETCD_STRICT_RECONFIG_CHECK="true"
#ETCD_ENABLE_V2="true"
```

在上面的配置文件中，主要需要修改的选项如下：

1. ETCD_LISTEN_PEER_URLS

该选项用来指定 etcd 节点监听的 URL，用于与其他分布式 etcd 节点通信，实现各个 etcd

节点的数据通信、交互、选举以及数据同步等功能。该 URL 采用协议、IP 和端口相组合的形式，可以是：

```
http://ip:port
```

或者：

```
https://ip:port
```

在本例中，节点的 IP 地址为 192.168.1.121，默认的端口为 2380。用户可以通过该选项同时指定多个 URL，各 URL 之间通过逗号隔开。

2．ETCD_LISTEN_CLIENT_URLS

该选项用于指定对外提供服务的地址，即 etcd API 的地址，etcd 客户端通过该 URL 访问 etcd 服务器。该选项同样采用协议、IP 地址和端口组合的形式，其默认端口为 2379。

3．ETCD_NAME

该选项用来指定 etcd 节点的名称，该名称用于在集群中标识本 etcd 节点。

4．ETCD_INITIAL_ADVERTISE_PEER_URLS

该选项指定节点同伴监听地址，这个值会告诉 etcd 集群中其他 etcd 节点，该地址用来在 etcd 集群中传递数据。

5．ETCD_ADVERTISE_CLIENT_URLS

该选项用来指定当前 etcd 节点对外公告的客户端监听地址，这个值会告诉集群中其他节点。

6．ETCD_INITIAL_CLUSTER

该选项列出当前 etcd 集群中的所有的 etcd 节点的节点通信地址。

在 Node1 节点上面执行以下命令，安装 Kubernetes 节点组件、etcd、flannel 以及 docker：

```
[root@localhost ~]# yum -y install kubernetes-node etcd flannel docker
```

安装完成之后，编辑/etc/etcd/etcd.conf 配置文件，修改内容如下：

```
#[Member]
#ETCD_CORS=""
ETCD_DATA_DIR="/var/lib/etcd/default.etcd"
#ETCD_WAL_DIR=""
ETCD_LISTEN_PEER_URLS="http://192.168.1.122:2380"
ETCD_LISTEN_CLIENT_URLS="http://192.178.1.122:2379,http://127.0.0.1:2379"
#ETCD_MAX_SNAPSHOTS="5"
#ETCD_MAX_WALS="5"
ETCD_NAME="etcd2"
#ETCD_SNAPSHOT_COUNT="100000"
```

```
#ETCD_HEARTBEAT_INTERVAL="100"
#ETCD_ELECTION_TIMEOUT="1000"
#ETCD_QUOTA_BACKEND_BYTES="0"
#ETCD_MAX_REQUEST_BYTES="1572864"
#ETCD_GRPC_KEEPALIVE_MIN_TIME="5s"
#ETCD_GRPC_KEEPALIVE_INTERVAL="2h0m0s"
#ETCD_GRPC_KEEPALIVE_TIMEOUT="20s"
#
#[Clustering]
ETCD_INITIAL_ADVERTISE_PEER_URLS="http://192.168.1.122:2380"
ETCD_ADVERTISE_CLIENT_URLS="http://192.168.1.122:2379"
#ETCD_DISCOVERY=""
#ETCD_DISCOVERY_FALLBACK="proxy"
#ETCD_DISCOVERY_PROXY=""
#ETCD_DISCOVERY_SRV=""
ETCD_INITIAL_CLUSTER="etcd1=http://192.168.1.121:2380,etcd2=http://192.16
8.1.122:2380,etcd3=http://192.168.1.123:2380"
#ETCD_INITIAL_CLUSTER_TOKEN="etcd-cluster"
#ETCD_INITIAL_CLUSTER_STATE="new"
#ETCD_STRICT_RECONFIG_CHECK="true"
#ETCD_ENABLE_V2="true"
#
#[Proxy]
#ETCD_PROXY="off"
#ETCD_PROXY_FAILURE_WAIT="5000"
#ETCD_PROXY_REFRESH_INTERVAL="30000"
#ETCD_PROXY_DIAL_TIMEOUT="1000"
#ETCD_PROXY_WRITE_TIMEOUT="5000"
#ETCD_PROXY_READ_TIMEOUT="0"
#
#[Security]
#ETCD_CERT_FILE=""
#ETCD_KEY_FILE=""
#ETCD_CLIENT_CERT_AUTH="false"
#ETCD_TRUSTED_CA_FILE=""
#ETCD_AUTO_TLS="false"
#ETCD_PEER_CERT_FILE=""
#ETCD_PEER_KEY_FILE=""
#ETCD_PEER_CLIENT_CERT_AUTH="false"
#ETCD_PEER_TRUSTED_CA_FILE=""
#ETCD_PEER_AUTO_TLS="false"
#
#[Logging]
```

```
#ETCD_DEBUG="false"
#ETCD_LOG_PACKAGE_LEVELS=""
#ETCD_LOG_OUTPUT="default"
#
#[Unsafe]
#ETCD_FORCE_NEW_CLUSTER="false"
#
#[Version]
#ETCD_VERSION="false"
#ETCD_AUTO_COMPACTION_RETENTION="0"
#
#[Profiling]
#ETCD_ENABLE_PPROF="false"
#ETCD_METRICS="basic"
#
#[Auth]
#ETCD_AUTH_TOKEN="simple"
```

在 Node2 节点上面执行同样的命令，安装相同的组件，只不过/etc/etcd/etcd.conf 文件中的内容与 Node1 节点稍有不同，需要修改的部分如下所示：

```
#[Member]
#ETCD_CORS=""
ETCD_DATA_DIR="/var/lib/etcd/default.etcd"
#ETCD_WAL_DIR=""
ETCD_LISTEN_PEER_URLS="http://192.168.1.123:2380"
ETCD_LISTEN_CLIENT_URLS="http://192.168.1.123:2379,http://127.0.0.1:2379"
#ETCD_MAX_SNAPSHOTS="5"
#ETCD_MAX_WALS="5"
ETCD_NAME="etcd3"
#ETCD_SNAPSHOT_COUNT="100000"
#ETCD_HEARTBEAT_INTERVAL="100"
#ETCD_ELECTION_TIMEOUT="1000"
#ETCD_QUOTA_BACKEND_BYTES="0"
#ETCD_MAX_REQUEST_BYTES="1572864"
#ETCD_GRPC_KEEPALIVE_MIN_TIME="5s"
#ETCD_GRPC_KEEPALIVE_INTERVAL="2h0m0s"
#ETCD_GRPC_KEEPALIVE_TIMEOUT="20s"
#
#[Clustering]
ETCD_INITIAL_ADVERTISE_PEER_URLS="http://192.168.1.123:2380"
ETCD_ADVERTISE_CLIENT_URLS="http://192.168.1.123:2379"
#ETCD_DISCOVERY=""
#ETCD_DISCOVERY_FALLBACK="proxy"
```

```
#ETCD_DISCOVERY_PROXY=""
#ETCD_DISCOVERY_SRV=""
#ETCD_INITIAL_CLUSTER="etcd1=http://192.168.1.121:2380,etcd2=http://192.168.1.122:2380,etcd3=http://192.168.1.123:2380"
#ETCD_INITIAL_CLUSTER_TOKEN="etcd-cluster"
#ETCD_INITIAL_CLUSTER_STATE="new"
#ETCD_STRICT_RECONFIG_CHECK="true"
#ETCD_ENABLE_V2="true"
…
```

配置完成之后，在 3 个节点上面分别执行以下命令，以启用和启动 etcd 服务：

```
[root@localhost etcd]# systemctl enable etcd
Created symlink from
/etc/systemd/system/multi-user.target.wants/etcd.service to
/usr/lib/systemd/system/etcd.service.
[root@localhost etcd]# systemctl start etcd
```

启动完成之后，通过以下命令查看 etcd 服务状态：

```
[root@localhost etcd]# systemctl status etcd
● etcd.service - Etcd Server
   Loaded: loaded (/usr/lib/systemd/system/etcd.service; enabled; vendor
preset: disabled)
   Active: active (running) since Fri 2019-03-08 08:15:52 CST; 11min ago
 Main PID: 28464 (etcd)
   CGroup: /system.slice/etcd.service
           └─28464 /usr/bin/etcd --name=etcd2
--data-dir=/var/lib/etcd/default.etcd
--listen-client-urls=http://192.168.1.122:2379,http://127.0.0.1:2379
```

如果上面的输出中的圆点是绿色的，并且 Active 的值为 active (running)，则表示服务启动成功。

etcd 提供的 etcdctl 命令可以查看 etcd 集群的健康状态，如下所示：

```
[root@localhost ~]# etcdctl cluster-health
member 1d661a1c0834462e is healthy: got healthy result from
http://192.168.1.123:2379
member cb2f5e732502a48c is healthy: got healthy result from
http://192.168.1.122:2379
member e4e48d7f05a9da0c is healthy: got healthy result from
http://192.168.1.121:2379
```

从上面的输出结果可以得知，集群中的 3 个 etcd 节点都是处于健康状态。

2.1.5　Master 节点配置

接下来介绍如何配置 Kubernetes 的 Master 节点。前面已经介绍过，在 Master 节点上面主要运行着 apiserver、controller-manager 以及 scheduler 等主要的服务进程。以上服务的配置文件都位于/etc/kubernetes 目录中。其中，通常需要配置的为 apiserver，其配置文件为/etc/kubernetes/apiserver。修改该文件内容，如下所示：

```
###
# kubernetes system config
#
# The following values are used to configure the kube-apiserver
#

# The address on the local server to listen to.
#KUBE_API_ADDRESS="--insecure-bind-address=127.0.0.1"
KUBE_API_ADDRESS="--address=0.0.0.0"
# The port on the local server to listen on.
KUBE_API_PORT="--port=8080"

# Port minions listen on
KUBELET_PORT="--kubelet-port=10250"

# Comma separated list of nodes in the etcd cluster
KUBE_ETCD_SERVERS="--etcd-servers=http://192.168.1.121:2379,http://192.168.1.122:2379,http://192.168.1.123:2379"

# Address range to use for services
KUBE_SERVICE_ADDRESSES="--service-cluster-ip-range=10.254.0.0/16"

# default admission control policies
KUBE_ADMISSION_CONTROL="--admission-control=NamespaceLifecycle,NamespaceExists,LimitRanger,ResourceQuota"

# Add your own!
KUBE_API_ARGS=""
```

KUBE_API_ADDRESS 选项表示 apiserver 进程绑定的 IP 地址，在本例中将其修改为--address=0.0.0.0，表示绑定本机所有的 IP 地址。KUBE_API_PORT 选项用来指定 apiserver 监听的端口。KUBELET_PORT 表示 kubelet 监听的服务端口。KUBE_ETCD_SERVERS 选项指定 etcd 集群中的每个节点的地址。KUBE_SERVICE_ADDRESSES 选项指定 Kubernetes 中的服务的 IP 地址范围。默认情况下，KUBE_ADMISSION_CONTRO 选项会包含 SecurityContextDeny 和 ServiceAccount，这 2 个值与权限有关，在测试的时候，可以将其去掉。

配置完成之后，使用以下命令启动 Master 节点上面的各项服务：

```
[root@localhost ~]# systemctl start kube-apiserver
[root@localhost ~]# systemctl start kube-controller-manager
[root@localhost ~]# systemctl start kube-scheduler
```

然后通过 systemctl 命令来确定各项服务是否启动成功，例如，可以使用下面的命令查看 apiserver 的状态：

```
[root@localhost ~]# systemctl status kube-apiserver
● kube-apiserver.service - Kubernetes API Server
   Loaded: loaded (/usr/lib/systemd/system/kube-apiserver.service; disabled;
vendor preset: disabled)
   Active: active (running) since Sat 2019-03-09 04:56:38 CST; 145ms ago
     Docs: https://github.com/GoogleCloudPlatform/kubernetes
 Main PID: 19080 (kube-apiserver)
   CGroup: /system.slice/kube-apiserver.service
           └─19080 /usr/bin/kube-apiserver --logtostderr=true --v=0
--etcd-servers=http://192.168.1.121:2379,http://192.168.1.122:2379,http://192
.168.1.123:2379 --address=0.0.0.0 --port=8080 --kubelet-port=...

Mar 09 04:56:38 localhost.localdomain kube-apiserver[19080]: I0309
04:56:38.339774   19080 config.go:562] Will report 192.168.1.121 as public IP
address.
Mar 09 04:56:38 localhost.localdomain kube-apiserver[19080]: W0309
04:56:38.340594   19080 handlers.go:50] Authentication is disabled
Mar 09 04:56:38 localhost.localdomain kube-apiserver[19080]: E0309
04:56:38.341144   19080 reflector.go:199]
k8s.io/kubernetes/plugin/pkg/admission/resourcequota/resource_access.go:83:
Failed to...tion refused
Mar 09 04:56:38 localhost.localdomain kube-apiserver[19080]: E0309
04:56:38.385934   19080 reflector.go:199]
pkg/controller/informers/factory.go:89: Failed to list *api.Namespace: Get
http://0.0...tion refused
Mar 09 04:56:38 localhost.localdomain kube-apiserver[19080]: E0309
04:56:38.390288   19080 reflector.go:199]
pkg/controller/informers/factory.go:89: Failed to list *api.LimitRange: Get
http://0....tion refused
Mar 09 04:56:38 localhost.localdomain kube-apiserver[19080]: [restful]
2019/03/09 04:56:38 log.go:30: [restful/swagger] listing is available at
https://192.168.1.121:6443/swaggerapi/
Mar 09 04:56:38 localhost.localdomain kube-apiserver[19080]: [restful]
2019/03/09 04:56:38 log.go:30: [restful/swagger]
https://192.168.1.121:6443/swaggerui/ is mapped to folder /swagger-ui/
```

```
    Mar 09 04:56:38 localhost.localdomain kube-apiserver[19080]: I0309
04:56:38.445352  19080 serve.go:95] Serving securely on 0.0.0.0:6443
    Mar 09 04:56:38 localhost.localdomain systemd[1]: Started Kubernetes API
Server.
    Mar 09 04:56:38 localhost.localdomain kube-apiserver[19080]: I0309
04:56:38.445442  19080 serve.go:109] Serving insecurely on 0.0.0.0:8080
    Hint: Some lines were ellipsized, use -l to show in full.
```

从上面的输出结果可以得知 apiserver 服务进程已经处于运行状态。

为了使得各项服务在 Linux 系统启动时自动启动，用户可以使用以下命令启用各项服务：

```
[root@localhost ~]# systemctl enable kube-apiserver
[root@localhost ~]# systemctl enable kube-controller-manager
[root@localhost ~]# systemctl enable kube-scheduler
```

Kubernetes 的 apiserver 提供的各个接口都是 RESTful 的，用户可以通过浏览器访问 Master 节点的 6443 端口，apiserver 会以 JSON 对象的形式返回各个 API 的地址，如图 2-2 所示。

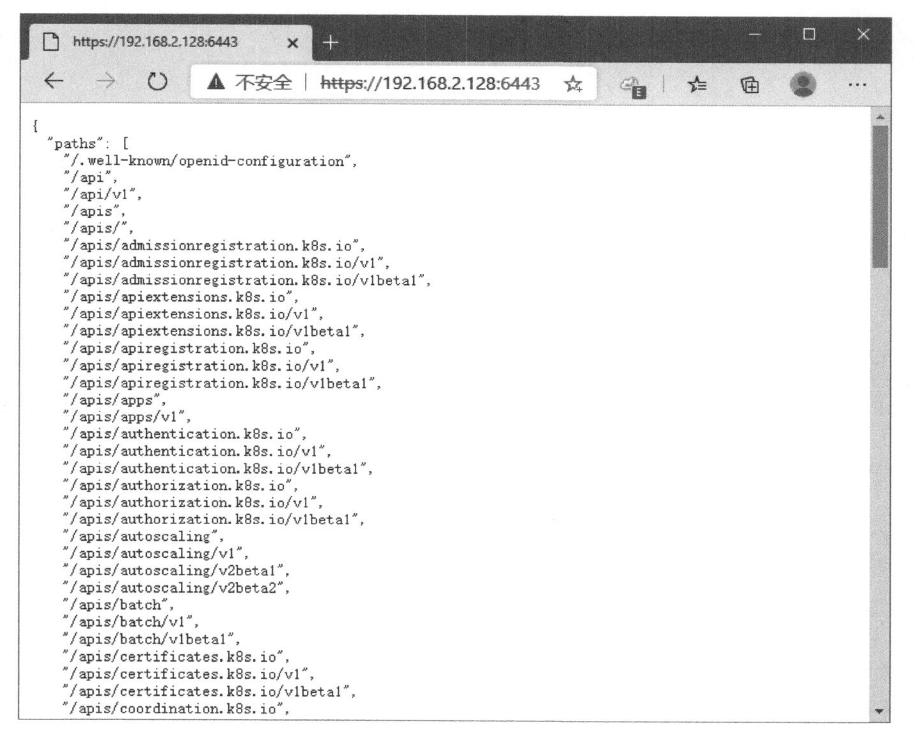

图 2-2　Kubernetes apiserver 提供的 API 接口

2.1.6　Node 节点配置

Node 节点上面主要运行 kube-proxy 以及 kubelet 等进程。用户需要修改的配置文件主要有 /etc/kubernetes/config、/etc/kubernetes/proxy 以及/etc/kubernetes/kubelet，这 3 个文件分别为 Kubernetes 全局配置文件、kube-proxy 配置文件以及 kubelet 配置文件。在所有的 Node 节点中，

这些配置文件大同小异，主要区别在于各个节点的 IP 地址会有所不同。下面以 Node1 为例，说明其配置方法。

首先修改/etc/kubernetes/config，主要修改 KUBE_MASTER 选项，指定 apiserver 的地址，如下所示：

```
###
# kubernetes system config
#
# The following values are used to configure various aspects of all
# kubernetes services, including
#
#   kube-apiserver.service
#   kube-controller-manager.service
#   kube-scheduler.service
#   kubelet.service
#   kube-proxy.service
# logging to stderr means we get it in the systemd journal
KUBE_LOGTOSTDERR="--logtostderr=true"

# journal message level, 0 is debug
KUBE_LOG_LEVEL="--v=0"

# Should this cluster be allowed to run privileged docker containers
KUBE_ALLOW_PRIV="--allow-privileged=false"

# How the controller-manager, scheduler, and proxy find the apiserver
KUBE_MASTER="--master=http://192.168.1.121:8080"
…
```

然后修改 kubelet 的配置文件，内容如下：

```
###
# kubernetes kubelet (minion) config

# The address for the info server to serve on (set to 0.0.0.0 or "" for all
interfaces)
KUBELET_ADDRESS="--address=127.0.0.1"

# The port for the info server to serve on
KUBELET_PORT="--port=10250"

# You may leave this blank to use the actual hostname
KUBELET_HOSTNAME="--hostname-override=192.168.1.122"
```

```
# location of the api-server
KUBELET_API_SERVER="--api-servers=http://192.168.1.121:8080"

# pod infrastructure container
KUBELET_POD_INFRA_CONTAINER="--pod-infra-container-image=registry.access.
redhat.com/rhel7/pod-infrastructure:latest"

# Add your own!
KUBELET_ARGS=""
```

其中，KUBELET_ADDRESS 指定 kubelet 绑定的 IP 地址，如果想要绑定本机所有的网络接口，可以将其指定为 0.0.0.0。KUBELET_PORT 指定 kubelet 监听的端口，KUBELET_HOSTNAME 指定本节点的主机名，该选项的值可以是主机名，也可以是本机的 IP 地址。在本例中，Node1 节点的 IP 地址为 192.168.1.122。KUBELET_API_SERVER 选项指定 apiserver 的地址。

注　意

如果 KUBELET_HOSTNAME 选项的值为主机名，则需要在 hosts 文件中配置主机名和 IP 地址的对应关系。

最后，修改/etc/kubernetes/proxy 文件，将其内容修改如下：

```
###
# kubernetes proxy config

# default config should be adequate

# Add your own!
KUBE_PROXY_ARGS="—bind-address=0.0.0.0"
```

配置完成之后，执行以下命令配置开机启动：

```
[root@localhost ~]# systemctl enable kube-proxy
Created symlink from /etc/systemd/system/multi-user.target.wants/
kube-proxy.service to /usr/lib/systemd/system/kube-proxy.service.
[root@localhost ~]# systemctl enable kubelet
Created symlink from /etc/systemd/system/multi-user.target.wants/
kubelet.service to /usr/lib/systemd/system/kubelet.service.
```

然后使用以下命令启动 kube-proxy 和 kubelet 服务：

```
[root@localhost ~]# systemctl start kube-proxy
[root@localhost ~]# systemctl start kubelet
```

按照上面的方法在 Node2 节点上面进行配置。在配置的过程中，注意要把相应的 IP 地址修改为 Node2 的 IP 地址 192.168.1.123。配置完成之后，分别启动 kube-proxy 和 kubelet。

最后再测试一下集群是否正常。在 Master 节点上面执行以下命令：

```
[root@localhost kubernetes]# kubectl get nodes
NAME                STATUS          AGE
192.168.1.122       Ready           19m
192.168.1.123       Ready           17s
```

如果上面的命令输出各 Node 节点，并且其状态为 Ready，则表示当前集群已经正常工作了。

2.1.7 配置网络

Flannel 是 Kubernetes 中常用的网络配置工具，用于配置第三层（网络层）网络结构。Flannel 需要在集群中的每台主机上运行一个名为 flanneld 的代理程序，负责从预配置地址空间中为每台主机分配一个网段。Flannel 直接使用 Kubernetes API 或 etcd 存储网络配置、分配的子网以及任何辅助数据。

在配置 Flannel 之前，用户需要预先设置分配给 Docker 网络的网段。在 Master 节点上面执行以下命令，在 etcd 中添加一个名为/atomic.io/network/config 的主键，通过该主键设置提供给 Docker 容器使用的网段以及子网。

```
[root@localhost ~]# etcdctl mk /atomic.io/network/config '{"Network":
"172.17.0.0/16", "SubnetMin": "172.17.1.0", "SubnetMax": "172.17.254.0"}'
```

然后在 Node1 和 Node2 这 2 个 Node 节点上面修改/etc/sysconfig/flanneld 配置文件，使其内容如下：

```
# Flanneld configuration options

# etcd url location.  Point this to the server where etcd runs
FLANNEL_ETCD_ENDPOINTS="http://192.168.1.121:2379,http://192.168.1.122:23
79,http://192.168.1.123:2379"

# etcd config key.  This is the configuration key that flannel queries
# For address range assignment
FLANNEL_ETCD_PREFIX="/atomic.io/network"

# Any additional options that you want to pass
FLANNEL_OPTIONS="--iface=ens33"
```

其中，FLANNEL_ETCD_ENDPOINTS 用来指定 etcd 集群的各个节点的地址。FLANNEL_ETCD_PREFIX 指定 etcd 中网络配置的主键，该主键要与前面设置的主键值完全一致。FLANNEL_OPTIONS 中的--iface 选项指定 Flannel 网络使用的网络接口。

设置完成之后，分别在 Node1 和 Node2 节点上面，通过以下 2 条命令启用和启动 flanneld：

```
[root@localhost ~]# systemctl enable flanneld
[root@localhost ~]# systemctl start flanneld
```

启动成功之后，通过 ip 命令查看系统中的网络接口，会发现多出一个名为 flannel0 的网络接口，如下所示：

```
[root@localhost ~]# ip address show | grep flannel
4: flannel0: <POINTOPOINT,MULTICAST,NOARP,UP,LOWER_UP> mtu 1472 qdisc
pfifo_fast state UNKNOWN group default qlen 500
    inet 172.17.99.0/16 scope global flannel0
```

从上面的输出可以得知，虚拟接口 flannel0 的 IP 地址是前面指定的 172.17.0.0/16。

此外，flannel 还生成了 2 个配置文件，分别是/run/flannel/subnet.env 和/run/flannel/docker。其中 subnet.env 的内容如下所示：

```
[root@localhost ~]# cat /run/flannel/subnet.env
FLANNEL_NETWORK=172.17.0.0/16
FLANNEL_SUBNET=172.17.93.1/24
FLANNEL_MTU=1472
FLANNEL_IPMASQ=false
```

docker 的内容如下所示：

```
[root@localhost ~]# cat /run/flannel/docker
DOCKER_OPT_BIP="--bip=172.17.93.1/24"
DOCKER_OPT_IPMASQ="--ip-masq=true"
DOCKER_OPT_MTU="--mtu=1472"
DOCKER_NETWORK_OPTIONS=" --bip=172.17.93.1/24 --ip-masq=true --mtu=1472"
```

关于 Flannel 的基本原理和详细使用方法，将在后面的章节中介绍。

2.2　通过二进制文件安装 Kubernetes

在前面一节中，介绍了通过软件包管理工具安装 Kubernetes。实际上，Kubernetes 还为多种软硬件平台提供了编译好的二进制文件。用户可以直接从官方网站上面下载这些二进制文件，然后稍加配置即可使用。本节将详细介绍二进制文件安装 Kubernetes 的方法。

2.2.1　安装前准备

在本例中，我们同样部署 3 个节点，其中一个为 Master 节点，另外两个为 Node 节点。其网络拓扑结构与图 2-1 完全相同。

1. 下载安装包

Kubernetes 二进制文件的下载地址为：https://github.com/kubernetes/kubernetes/releases，其中，最新的正式版为 v1.19.5，如图 2-3 所示。

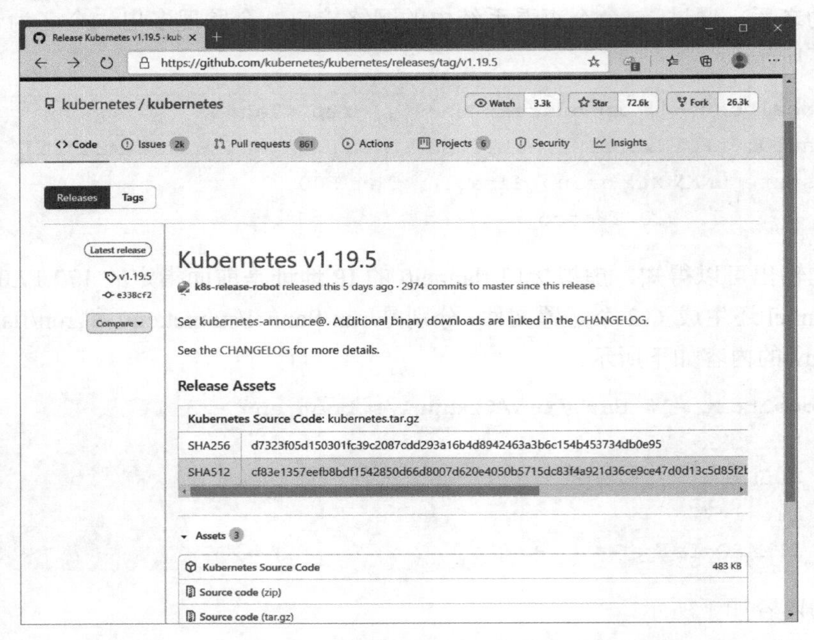

图 2-3 Kubernetes 二进制文件下载

如果用户想要下载其他软硬件平台的二进制文件，可以单击其中的 the CHANGELOG 链接，跳转到另外一个下载页面，如图 2-4 所示。从图中可以得知，Kubernetes 为每个版本都提供了许多平台的二进制软件包。

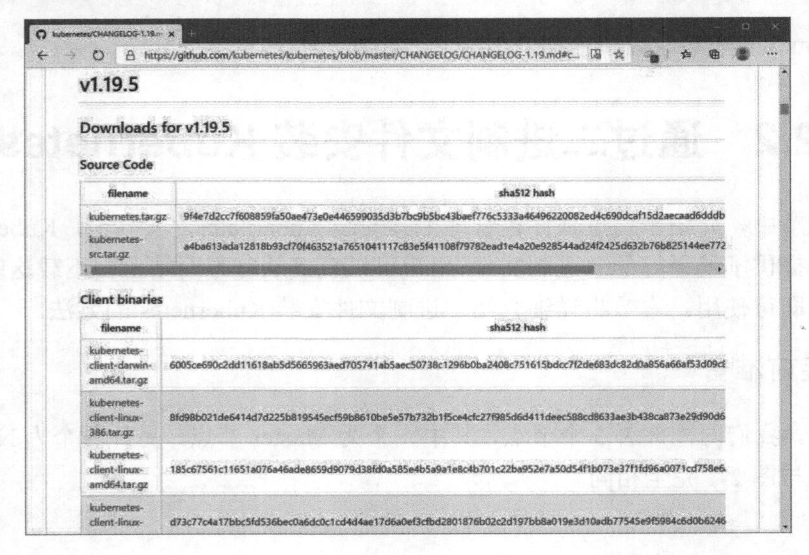

图 2-4 v1.19.5 下载页面

从网页上面可以得知，Kubernetes将二进制包分为Client Binaries、Server Binaries以及Node Binaries，分别对应着客户端二进制包、Master 节点二进制包以及 Node 节点二进制包。每个文件又分为 Darwin、Linux 以及 Windows 等操作系统平台。除此之外，还有 386、amd64、ppc64 以及 s390 等硬件平台。

在本例中，我们将在64位的CentOS上面安装，所以需要下载kubernetes-server-linux-amd64.tar.gz、kubernetes-node-linux-amd64.tar.gz 以及 kubernetes-client-linux-amd64.tar.gz 。实际上 kubernetes-client-linux-amd64.tar.gz 只包含一个文件 kubectl，而这个文件已经包含在其他的两个压缩文件中，所以在安装的时候，可以不单独下载该文件。

通常情况下，用户应该将第三方的软件包保存在/opt 目录中。所以，首先执行以下命令，切换到该目录中：

```
[root@localhost ~]# cd /opt/
```

然后执行以下两条命令，分别下载 Master 和 Node 节点的二进制文件。对于 Master 节点，用户只需要下载 kubernetes-server-linux-amd64.tar.gz 即可，对于 Node 节点，用户只需要下载 kubernetes-node-linux-amd64.tar.gz 文件。

```
    [root@localhost opt]# wget https://dl.k8s.io/v1.19.5/kubernetes-server-
linux-amd64.tar.gz
    [root@localhost opt]# wget https://dl.k8s.io/v1.19.5/kubernetes-node-linux-
amd64.tar.gz
```

etcd 的二进制文件并没有包含在 Kubernetes 的二进制压缩包中，用户需要单独下载。与 Kubernetes 一样，etcd 也提供了多种软硬件平台的二进制文件，其网址为：https://github.com/etcd-io/etcd/releases。在所有节点上面，执行以下命令下载 etcd 的二进制文件：

```
    [root@localhost opt]# wget https://github.com/etcd-io/etcd/releases/
download/v3.4.14/etcd-v3.4.14-linux-amd64.tar.gz
```

2. 关闭防火墙和 SELinux

在所有节点上面执行以下命令设置防火墙和 SELinux：

```
[root@localhost opt]# systemctl stop firewalld && systemctl disable firewalld
[root@localhost opt]# setenforce 0
```

编辑/etc/selinux/config 文件，将 SELINUX 设置为 disabled，如下所示：

```
SELINUX=disabled
```

3. 禁用交换分区

在所有节点上面使用以下命令禁用 CentOS 的交换分区：

```
[root@localhost opt]# swapoff -a && sysctl -w vm.swappiness=0
vm.swappiness = 0
```

然后修改/etc/fstab 文件，将交换分区对应的文件系统注释掉，如下所示：

```
#/dev/mapper/centos-swap swap                    swap    defaults        0 0
```

4. 设置 Docker 所需要的网络参数

由于在本例中，Docker 只安装在 Node 节点上，因此在所有 Node 节点上修改/etc/sysctl.d/k8s.conf，增加以下行：

```
net.ipv4.ip_forward = 1
```

然后使用以下命令使得以上设置生效：

```
[root@localhost ~]# sysctl -p /etc/sysctl.d/k8s.conf
```

5. 配置 Docker 的 yum 安装源

默认情况下，CentOS 没有配置 Docker 的 yum 安装源，读者可以使用以下命令自行添加：

```
    [root@localhost ~]# yum-config-manager --add-repo
https://download.docker.com/linux/centos/docker-ce.repo
```

在上面的命令中，yum-config-manager 用来管理和配置 CentOS 的软件仓库。如果当前系统中没有该命令，那么可以使用以下命令安装：

```
[root@localhost ~]# yum install yum-utils
```

6. 安装 Docker

设置好软件源之后，就可以直接使用 yum 命令来安装 Docker CE 了，命令如下：

```
[root@localhost ~]# yum -y install docker-ce
```

安装完成之后，使用以下命令启用和启动 Docker 服务：

```
[root@localhost ~]# systemctl enable docker
[root@localhost ~]# systemctl start docker
```

7. 创建安装目录

在本例中，我们计划将 Kubernetes 安装在/k8s 目录中，所以使用以下命令分别创建几个相关的目录：

```
    [root@localhost ~]# mkdir -p /k8s/etcd/bin k8s/etcd/cfg
    [root@localhost ~]# mkdir -p /k8s/kubernetes/bin /k8s/kubernetes/cfg
/k8s/kubernetes/ssl
```

2.2.2 部署 etcd

使用以下命令解压前面下载的 etcd-v3.4.14-linux-amd64.tar.gz：

```
[root@localhost opt]# tar -xvf etcd-v3.4.14-linux-amd64.tar.gz
```

然后将解压后的目录中的 etcd 和 etcdctl 这两个文件复制到/k8s/etcd/bin 目录中：

```
[root@localhost opt]# cp etcd-v3.4.14-linux-amd64/etcd
etcd-v3.4.14-linux-amd64/etcdctl /k8s/etcd/bin/
```

接下来配置 etcd。由于我们要在 3 个节点上都部署 etcd，形成一个 etcd 集群，因此接下来需要分别在 3 个主机上创建 etcd 的配置文件。跟前一节中介绍的一样，所有的 etcd 节点的配置文件基本相同，区别在于绑定的 IP 地址有所不同。

首先是 Master 节点，使用以下命令创建配置文件：

```
[root@localhost ~]# vi /k8s/etcd/cfg/etcd
```

其内容如下：

```
#[Member]
ETCD_NAME="etcd01"
ETCD_DATA_DIR="/var/lib/etcd/default.etcd"
ETCD_LISTEN_PEER_URLS="http://192.168.1.121:2380"
ETCD_LISTEN_CLIENT_URLS="http://192.168.1.121:2379"

#[Clustering]
ETCD_INITIAL_ADVERTISE_PEER_URLS="http://192.168.1.121:2380"
ETCD_ADVERTISE_CLIENT_URLS="http://192.168.1.121:2379"
ETCD_INITIAL_CLUSTER="etcd01=http://192.168.1.121:2380,etcd02=http://192.
168.1.122:2380,etcd03=http://192.168.1.123:2380"
ETCD_INITIAL_CLUSTER_TOKEN="etcd-cluster"
ETCD_INITIAL_CLUSTER_STATE="new"
```

接下来是 Node1，创建配置文件的命令如下：

```
[root@localhost ~]# vi /k8s/etcd/cfg/etcd
```

其内容如下：

```
#[Member]
ETCD_NAME="etcd02"
ETCD_DATA_DIR="/var/lib/etcd/default.etcd"
ETCD_LISTEN_PEER_URLS="http://192.168.1.122:2380"
ETCD_LISTEN_CLIENT_URLS="http://192.168.1.122:2379"

#[Clustering]
ETCD_INITIAL_ADVERTISE_PEER_URLS="http://192.168.1.122:2380"
ETCD_ADVERTISE_CLIENT_URLS="http://192.168.1.122:2379"
ETCD_INITIAL_CLUSTER="etcd01=http://192.168.1.121:2380,etcd02=http://192.
168.1.122:2380,etcd03=http://192.168.1.123:2380"
ETCD_INITIAL_CLUSTER_TOKEN="etcd-cluster"
ETCD_INITIAL_CLUSTER_STATE="new"
```

接下来是 Node2，创建配置文件的命令如下：

```
[root@localhost opt]# vi /k8s/etcd/cfg/etcd
```

其内容如下：

```
#[Member]
ETCD_NAME="etcd03"
ETCD_DATA_DIR="/var/lib/etcd/default.etcd"
ETCD_LISTEN_PEER_URLS="http://192.168.1.123:2380"
ETCD_LISTEN_CLIENT_URLS="http://192.168.1.123:2379"

#[Clustering]
ETCD_INITIAL_ADVERTISE_PEER_URLS="http://192.168.1.123:2380"
ETCD_ADVERTISE_CLIENT_URLS="http://192.168.1.123:2379"
ETCD_INITIAL_CLUSTER="etcd01=http://192.168.1.121:2380,etcd02=http://192.
168.1.122:2380,etcd03=http://192.168.1.123:2380"
ETCD_INITIAL_CLUSTER_TOKEN="etcd-cluster"
ETCD_INITIAL_CLUSTER_STATE="new"
```

然后创建 etcd 的系统服务单元文件，由于在 3 个节点中，etcd 的安装位置完全相同，因此这 3 个节点的 etcd 的系统服务单元文件完全相同。在所有的节点上执行以下命令创建该文件：

```
[root@localhost ~]# vi /lib/systemd/system/etcd.service
```

其内容如下：

```
[Unit]
Description=Etcd Server
After=network.target
After=network-online.target
Wants=network-online.target

[Service]
Type=notify
EnvironmentFile=/k8s/etcd/cfg/etcd
ExecStart=/k8s/etcd/bin/etcd \
--name=${ETCD_NAME} \
--data-dir=${ETCD_DATA_DIR} \
--listen-peer-urls=${ETCD_LISTEN_PEER_URLS} \
--listen-client-urls=${ETCD_LISTEN_CLIENT_URLS},http://127.0.0.1:2379 \
--advertise-client-urls=${ETCD_ADVERTISE_CLIENT_URLS} \
--initial-advertise-peer-urls=${ETCD_INITIAL_ADVERTISE_PEER_URLS} \
--initial-cluster=${ETCD_INITIAL_CLUSTER} \
--initial-cluster-token=${ETCD_INITIAL_CLUSTER_TOKEN} \
```

```
--initial-cluster-state=new
Restart=on-failure
LimitNOFILE=65536

[Install]
WantedBy=multi-user.target
```

最后，在 3 个节点上使用以下命令启动 etcd 服务：

```
[root@localhost ~]# systemctl enable etcd
[root@localhost ~]# systemctl start etcd
```

配置完成之后，用户可以使用以下命令验证集群是否正常运行：

```
[root@localhost bin]# /k8s/etcd/bin/etcdctl cluster-health
member 1d661a1c0834462e is healthy: got healthy result from
http://192.168.1.123:2379
member cb2f5e732502a48c is healthy: got healthy result from
http://192.168.1.122:2379
member e4e48d7f05a9da0c is healthy: got healthy result from
http://192.168.1.121:2379
cluster is healthy
```

从上面的输出结果可知，3 个节点都正常运行，整个集群也是正常的。

注　意
启动 etcd 集群时需要同时启动所有的节点。

2.2.3　部署 flannel 网络

分别在两个 Node 节点上从以下网址下载 flannel 的二进制文件：

https://github.com/coreos/flannel/releases

下载后的文件名为 flannel-v0.13.0-linux-amd64.tar.gz。
使用以下命令解压该文件：

```
[root@localhost opt]# tar zxvf flannel-v0.13.0-linux-amd64.tar.gz
```

然后将解压得到的 flanneld 和 mk-docker-opts.sh 这两个文件复制到/k8s/Kubernetes/bin 目录中：

```
[root@localhost opt]# cp flanneld mk-docker-opts.sh /k8s/kubernetes/bin/
```

创建 flannel 配置文件/k8s/kubernetes/cfg/flanneld，其内容如下：

```
FLANNEL_OPTIONS="--etcd-endpoints=http://192.168.1.121:2379,http://192.16
8.1.122:2379,http://192.168.1.123:2379"
```

在 etcd 集群中写入 Pod 的网络信息，如下所示：

```
    [root@localhost bin]# /k8s/etcd/bin/etcdctl set /coreos.com/network/config
'{ "Network": "172.18.0.0/16", "Backend": {"Type": "vxlan"}}'
```

创建 flannel 的系统服务单元文件/lib/systemd/system/flanneld.service，其内容如下：

```
    [Unit]
    Description=Flanneld overlay address etcd agent
    After=network-online.target network.target
    Before=docker.service

    [Service]
    Type=notify
    EnvironmentFile=/k8s/kubernetes/cfg/flanneld
    ExecStart=/k8s/kubernetes/bin/flanneld --ip-masq $FLANNEL_OPTIONS
    ExecStartPost=/k8s/kubernetes/bin/mk-docker-opts.sh -k
DOCKER_NETWORK_OPTIONS -d /run/flannel/subnet.env
    Restart=on-failure

    [Install]
    WantedBy=multi-user.target
```

最后启用并启动 flannel 服务：

```
    [root@localhost ~]# systemctl enable flanneld
    [root@localhost ~]# systemctl start flanneld
```

2.2.4 部署 Master 节点

前面已经介绍过，Kubernetes 的 Master 节点主要运行 kube-apiserver、kube-scheduler、kube-controller-manager 等组件。下面分别介绍这些组件的配置方法。

首先将前面下载的 kubernetes-server-linux-amd64.tar.gz 文件解压，然后将解压得到的目录中的 server/bin 下面的 kube-apiserver、kube-controller-manager、kube-scheduler 以及 kubectl 这 4 个文件复制到/k8s/kubernetes/bin 目录下：

```
    [root@localhost opt]# cd kubernetes/server/bin/
    [root@localhost bin]# cp kube-apiserver kube-controller-manager
kube-scheduler kubectl/k8s/kubernetes/bin/
```

1. 配置 kube-apiserver

创建 kube-apiserver 的配置文件/k8s/kubernetes/cfg/kube-apiserver，其内容如下：

```
    KUBE_APISERVER_OPTS="--logtostderr=true \
    --v=4 \
    --etcd-servers=http://192.168.1.121:2379,http://192.168.1.122:2379,http:/
/192.168.1.123:2379 \
```

```
    --address=0.0.0.0 \
    --port=8080 \
    --advertise-address=192.168.1.121 \
    --allow-privileged=true \
    --service-cluster-ip-range=10.0.0.0/24 \
    --enable-admission-plugins=NamespaceLifecycle,NamespaceExists,LimitRanger,
ResourceQuota"
```

然后创建 kube-apiserver 的系统服务单元文件/lib/systemd/system/kube-apiserver.service，内容如下：

```
[Unit]
Description=Kubernetes API Server
Documentation=https://github.com/kubernetes/kubernetes

[Service]
EnvironmentFile=-/k8s/kubernetes/cfg/kube-apiserver
ExecStart=/k8s/kubernetes/bin/kube-apiserver $KUBE_APISERVER_OPTS
Restart=on-failure

[Install]
WantedBy=multi-user.target
```

执行以下命令启动 kube-apiserver 服务：

```
[root@localhost ~]# systemctl enable kube-apiserver
[root@localhost ~]# systemctl start kueb-apiserver
```

查看服务状态：

```
[root@localhost ~]# systemctl status kube-apiserver
● kube-apiserver.service - Kubernetes API Server
   Loaded: loaded (/usr/lib/systemd/system/kube-apiserver.service; enabled;
vendor preset: disabled)
   Active: active (running) since Mon 2019-03-11 02:58:28 CST; 6min ago
     Docs: https://github.com/kubernetes/kubernetes
 Main PID: 25322 (kube-apiserver)
   CGroup: /system.slice/kube-apiserver.service
           └─25322 /k8s/kubernetes/bin/kube-apiserver --logtostderr=true
--v=4 --etcd-servers=http://192.168.1.121:2379,http://192.168.1.122:2379,
http://192.168.1.123:2379 --address=0.0.0.0 --port=8080 --ad...
```

2. 配置 kube-scheduler

创建 kube-scheduler 配置文件/k8s/kubernetes/cfg/kube-scheduler，其内容如下：

```
KUBE_SCHEDULER_OPTS="--logtostderr=true --v=4 --master=127.0.0.1:8080
--leader-elect"
```

创建 kube-scheduler 系统服务单元文件/lib/systemd/system/kube-scheduler.service，内容如下：

```
[Unit]
Description=Kubernetes Scheduler
Documentation=https://github.com/kubernetes/kubernetes

[Service]
EnvironmentFile=-/k8s/kubernetes/cfg/kube-scheduler
ExecStart=/k8s/kubernetes/bin/kube-scheduler $KUBE_SCHEDULER_OPTS
Restart=on-failure

[Install]
WantedBy=multi-user.target
```

启动服务，并查看服务状态：

```
[root@localhost ~]# systemctl enable kube-scheduler
Created symlink from
/etc/systemd/system/multi-user.target.wants/kube-scheduler.service to
/usr/lib/systemd/system/kube-scheduler.service.
[root@localhost ~]# systemctl start kube-scheduler
[root@localhost ~]# systemctl status kube-scheduler
● kube-scheduler.service - Kubernetes Scheduler
   Loaded: loaded (/usr/lib/systemd/system/kube-scheduler.service; enabled;
vendor preset: disabled)
   Active: active (running) since Mon 2019-03-11 03:07:51 CST; 4s ago
     Docs: https://github.com/kubernetes/kubernetes
 Main PID: 25893 (kube-scheduler)
   CGroup: /system.slice/kube-scheduler.service
           └─25893 /k8s/kubernetes/bin/kube-scheduler --logtostderr=true
--v=4 --master=127.0.0.1:8080 --leader-elect
```

3. 配置 kube-controller-manager

创建配置文件/k8s/kubernetes/cfg/kube-controller-manager，内容如下：

```
KUBE_CONTROLLER_MANAGER_OPTS="--logtostderr=true \
--v=4 \
--master=127.0.0.1:8080 \
--leader-elect=true \
--address=127.0.0.1 \
--service-cluster-ip-range=10.0.0.0/24 \
--cluster-name=kubernetes"
```

创建 kube-controller-manager 系统服务单元文件/lib/systemd/system/kube-controller-manager.service，内容如下;

```
    [Unit]
    Description=Kubernetes Controller Manager
    Documentation=https://github.com/kubernetes/kubernetes

    [Service]
    EnvironmentFile=-/k8s/kubernetes/cfg/kube-controller-manager
    ExecStart=/k8s/kubernetes/bin/kube-controller-manager
$KUBE_CONTROLLER_MANAGER_OPTS
    Restart=on-failure

    [Install]
    WantedBy=multi-user.target
```

启动服务，并检查状态：

```
    [root@localhost ~]# systemctl enable kube-controller-manager
    Created symlink from
/etc/systemd/system/multi-user.target.wants/kube-controller-manager.service to
/usr/lib/systemd/system/kube-controller-manager.service.
    [root@localhost ~]# systemctl restart kube-controller-manager
    [root@localhost ~]# systemctl status kube-controller-manager
    ● kube-controller-manager.service - Kubernetes Controller Manager
      Loaded: loaded (/usr/lib/systemd/system/kube-controller-manager.service;
enabled; vendor preset: disabled)
      Active: active (running) since Mon 2019-03-11 03:14:01 CST; 7s ago
        Docs: https://github.com/kubernetes/kubernetes
     Main PID: 26461 (kube-controller)
      CGroup: /system.slice/kube-controller-manager.service
              └─26461 /k8s/kubernetes/bin/kube-controller-manager
--logtostderr=true --v=4 --master=127.0.0.1:8080 --leader-elect=true
--address=127.0.0.1 --service-cluster-ip-range=10.0.0.0/24 --cluster-name=…
```

可以通过 kubectl 命令查看集群中的各个组件的状态，如下所示：

```
    [root@localhost ~]# /k8s/kubernetes/bin/kubectl get cs
    NAME                 STATUS    MESSAGE              ERROR
    scheduler            Healthy   ok
    controller-manager       Healthy    ok
    etcd-0               Healthy   {"health":"true"}
    etcd-1               Healthy   {"health":"true"}
    etcd-2               Healthy   {"health":"true"}
```

2.2.5 部署 Node 节点

Node 节点上主要运行 kubelet 和 kube-proxy 等组件。其中 kubelet 运行在每个工作节点上，接收 kube-apiserver 发送的请求，管理 Pod 容器，执行交互式命令。kubelet 启动时自动向 kube-apiserver 注册节点信息。kube-proxy 监听 kube-apiserver 中服务和端点的变化情况，创建路由规则来进行服务负载均衡。

在 2 个 Node 节点上解压前面下载的 kubernetes-node-linux-amd64.tar.gz 文件，然后将得到的目录中的 node/bin 目录中的文件复制到/k8s/Kubernetes/bin 目录中，如下所示：

```
[root@localhost opt]# tar zxvf kubernetes-node-linux-amd64.tar.gz
[root@localhost opt]# cp kubernetes/node/bin/* /k8s/kubernetes/bin/
```

由于所有的 Node 节点的配置基本一致，不同之处在于各个节点的 IP 地址不同，因此下面以 Node1 为例来介绍 Node 节点的部署。

1. 配置 kubelet

首先创建 kubelet 参数配置模板文件/k8s/kubernetes/cfg/kubelet.config，代码如下：

```
apiVersion: v1
clusters:
- cluster:
    server: http://192.168.1.121:8080
  name: kubernetes
contexts:
- context:
    cluster: kubernetes
    user: kubelet-bootstrap
  name: default
current-context: default
kind: Config
```

然后创建 kubelet 配置文件/k8s/kubernetes/cfg/kubelet，其内容如下：

```
KUBELET_OPTS="--register-node=true \
--allow-privileged=true \
--hostname-override=192.168.1.122 \
--kubeconfig=/k8s/kubernetes/cfg/kubelet.config \
--cluster-dns=10.254.0.2 \
--cluster-domain=cluster.local \
--pod-infra-container-image=registry.access.redhat.com/rhel7/pod-infrastr
ucture:latest \
--logtostderr=true"
```

接着创建 kubelet 系统服务单元文件，如下所示：

```
[Unit]
Description=Kubernetes Kubelet
After=docker.service
Requires=docker.service

[Service]
EnvironmentFile=/k8s/kubernetes/cfg/kubelet
ExecStart=/k8s/kubernetes/bin/kubelet $KUBELET_OPTS
Restart=on-failure
KillMode=process
[Install]
WantedBy=multi-user.target
```

最后启动 kubelet 服务，命令如下：

```
[root@localhost ~]# systemctl enable kubelet
[root@localhost ~]# systemctl start kubelet
```

2．部署 kube-proxy

正如前面介绍的一样，kube-proxy 运行在所有的 Node 节点上。Kube-proxy 监听 apiserver 中的服务和端点的变化情况，创建路由规则来进行服务负载均衡。

创建 kube-proxy 配置文件/k8s/kubernetes/cfg/kubelet-proxy，内容如下：

```
KUBE_PROXY_OPTS="--logtostderr=true \
--hostname-override=192.168.1.122 \
--master=http://192.168.1.121:8080"
```

创建 kube-proxy 系统服务单元文件/lib/systemd/system/kube-proxy.service，内容如下：

```
[Unit]
Description=Kubernetes Proxy
After=network.target
[Service]
EnvironmentFile=/k8s/kubernetes/cfg/kube-proxy
ExecStart=/k8s/kubernetes/bin/kube-proxy $KUBE_PROXY_OPTS
Restart=on-failure

[Install]
wantedBy=multi-user.target
```

然后启用和启动 kube-proxy，命令如下：

```
[root@localhost ~]# systemctl enable kube-proxy
[root@localhost ~]# systemctl start kube-proxy
```

部署完成之后，在 Master 节点上通过 kubectl 命令查看节点和组件状态，如下所示：

```
[root@localhost ~]# kubectl get nodes,cs
NAME                    STATUS     ROLES      AGE        VERSION
node/192.168.1.122      Ready      <none>     24h        v1.13.4
node/192.168.1.123      Ready      <none>     13s        v1.13.4

NAME                              STATUS     MESSAGE              ERROR
componentstatus/controller-manager   Healthy    ok
componentstatus/scheduler            Healthy    ok
componentstatus/etcd-2             Healthy    {"health":"true"}
componentstatus/etcd-0             Healthy    {"health":"true"}
componentstatus/etcd-1             Healthy    {"health":"true"}
```

注　意

在部署 Kubelet 时，一定要禁用主机的交换分区，否则会导致 Kubelet 启动失败。

2.3 通过源代码安装 Kubernetes

Kubernetes 是一个非常棒的容器集群管理系统。通常情况下，我们并不需要修改 Kubernetes 的代码即可直接使用。但如果我们在环境中发现了某个问题或者缺陷，或者按照特定业务需求需要修改 Kubernetes 代码时，为了让修改生效，那么就需要编译 Kubernetes 的代码了。本节将详细介绍 Kubernetes 的源代码安装方法。

大致上，Kubernetes 的源代码编译有两种方式，其中一种方式为本地二进制文件编译，直接将源代码编译成本地二进制可执行文件；另外一种方式为 Docker 镜像编译 Kubernetes，可以编译出各核心组件的二进制文件以及对应的镜像文件。我们首先介绍本地二进制文件编译，然后再介绍 Docker 镜像编译。

2.3.1 本地二进制文件编译

本地二进制可执行文件编译需要安装 Go 运行环境，命令如下：

```
[root@localhost opt]# wget -c https://dl.google.com/go/go1.11.4.linux-amd64.tar.gz -P /opt/
```

上面的命令将 Go 可执行文件下载到/opt 目录中。下载完成之后，进入到该目录，命令如下：

```
[root@localhost ~]# cd /opt/
```

接下来将 Go 软件包释放到/usr/local 目录下，命令如下：

```
[root@localhost opt]# tar -C /usr/local -xzf go1.11.4.linux-amd64.tar.gz
```

最后配置 PATH 变量，将 Go 可执行文件的路径加入进去：

```
[root@localhost opt]# echo "export PATH=$PATH:/usr/local/go/bin" >>
/etc/profile && source /etc/profile
```

准备好编译 Kubernetes 所需要的依赖之后，就可以下载 Kubernetes 的源代码了。通常情况下，用户应该为 Kubernetes 的源代码创建一个专门的目录。在本例中，在/opt 目录中创建一个名为 k8s 的目录，命令如下：

```
[root@localhost ~]# mkdir /opt/k8s
```

切换到该目录之后，通过 git 命令将 Kubernetes 的源代码克隆到本地，并且指定版本为 1.13，如下所示：

```
[root@localhost k8s]# git clone https://github.com/kubernetes/kubernetes -b
release-1.19
```

git 命令会在当前目录中自动创建一个名为 kubernetes 的目录，所有的源代码都在改目录中。切换到该目录，然后执行以下命令进行编译：

```
[root@localhost kubernetes]# make all
```

编译完成之后，如果没有出错，则将生成 Kubernetes 的各种可执行文件，文件位于_output/bin 目录中，如下所示：

```
[root@localhost kubernetes]# ll _output/bin/
total 1648508
-rwxr-xr-x 1 root root 39744960 Mar 16 19:59
apiextensions-apiserver
…
-rwxr-xr-x 1 root root 177479616 Mar 16 19:59 hyperkube
-rwxr-xr-x 1 root root 36337760 Mar 16 19:59 kubeadm
-rwxr-xr-x 1 root root 138464480 Mar 16 19:59
kube-apiserver
-rwxr-xr-x 1 root root 103737216 Mar 16 19:59
kube-controller-manager
-rwxr-xr-x 1 root root 39161280 Mar 16 19:59 kubectl
-rwxr-xr-x 1 root root 112851008 Mar 16 19:59 kubelet
-rwxr-xr-x 1 root root 110721528 Mar 16 19:59 kubemark
-rwxr-xr-x 1 root root 34746688 Mar 16 19:59 kube-proxy
-rwxr-xr-x 1 root root 37202144 Mar 16 19:59 kube-scheduler
-rwxr-xr-x 1 root root 4966976 Mar 16 19:59 linkcheck
-rwxr-xr-x 1 root root 1595200 Mar 16 19:59 mounter
-rwxr-xr-x 1 root root 10363744 Mar 16 19:16 openapi-gen
```

从上面的输出可以看到，通过编译，已经生成了 kubeadmin、kube-apiserver、kubectl 以及 kubelet 等前面已经介绍过的可执行文件。

接下来，用户就可以按照 2.2 节介绍的方法安装、部署 Kubernetes 了。

注　意

源代码编译 Kubernetes 时需要较大的内存，建议主机拥有 16GB 以上的物理内存，否则会出现内存溢出而导致编译失败。

2.3.2　Docker 镜像编译

Docker 镜像编译比较简单，用户只要先将 Kubernetes 的源代码克隆到本地，然后执行 make quick-release 命令即可：

```
$ git clone https://github.com/kubernetes/kubernetes
$ cd kubernetes
$ make quick-release
```

第 3 章

Kubernetes命令行工具

Kubernetes 提供了许多功能强大的命令行工具。对于系统管理员来说，绝大部分的管理工作都是通过命令行工具来完成的。因此，掌握好命令行工具，对于系统管理员来说非常有必要。本章将介绍 Kubernetes 提供的主要命令行工具。

本章涉及的知识点有：

- kubectl 的使用方法：主要介绍如何通过 kubectl 对 Kubernetes 中的各种资源对象进行管理。
- kubeadm 的使用方法：介绍如何通过 kubeadm 安装、部署 Kubernetes 集群。

3.1 kubectl 的使用方法

kubekubectl 是 Kubernetes 集群的命令行工具，通过 kubectl 能够对集群本身进行管理，并能够在集群上进行容器化应用的安装部署。可以说，kubectl 是 Kubernetes 集群的最重要的工具，是整个集群的大管家。本节将详细介绍 kubectl 的使用方法，以便于后面章节的学习。

3.1.1 kubectl 用法概述

kubectl 是一个 Kubernetes 的客户端工具。在前面介绍 Kubernetes 的安装方法时，kubectl 通常会在 Kubernetes 的 Master 节点的安装过程中被安装。实际上，kubectl 可以单独安装和运行，它可以安装在任意一台 Linux、Windows 或者 Mac 计算机上，只要这台计算机能够连接 Master 节点，就可以通过它来管理 Kubernetes 集群。

例如，在 Mac OS 上面，用户可以通过以下命令下载最新版本的 kubectl：

```
    curl -LO https://storage.googleapis.com/kubernetes-release/release/$(curl
-s https://storage.googleapis.com/kubernetes-release/release/stable.txt)/
bin/darwin/amd64/kubectl
```

如果要下载特定版本的 kubectl，则只需要将其中的$(curl -s https://storage.googleapis.com/kubernetes-release/release/stable.txt)替换为指定的版本号即可。例如，用户想要下载 v1.19.5，可以通过以下链接来下载：

https://storage.googleapis.com/kubernetes-release/release/v1.19.5/bin/darwin/amd64/kubectl

下载后得到的是二进制的 kubectl 可执行文件，用户需要通过修改其权限，使其可以执行，命令如下：

```
chmod +x kubectl
```

chmod 命令用来修改文件的模式位，+x 表示增加可执行权限。

除此之外，Mac OS 用户还可以使用以下命令安装 kubectl：

```
sudo port selfupdate
sudo port install kubectl
```

如果在 Linux 上面下载 kubectl，同样可以使用 curl 命令。例如，下面的命令下载 v1.19.5 版本的 kubectl：

```
    [root@localhost ~]#  curl -LO https://storage.googleapis.com/
kubernetes-release/release/v1.19.5/bin/linux/amd64/kubectl
```

除了 curl 命令之外，在 Linux 系统中，用户还可以通过 wget 命令下载 kubectl，如下所示：

```
    [root@localhost ~]# wget https://storage.googleapis.com/kubernetes-release/
release/v1.19.5/bin/linux/amd64/kubectl
```

对于 Windows 用户来说，Kubernetes 也提供了相应的版本。例如，用户可以通过以下网址下载 v1.19.5 的 64 位的 Windows 版本的 kubectl：

https://storage.googleapis.com/kubernetes-release/release/v1.19.5/bin/windows/amd64/kubectl.exe

kubectl 的基本语法如下：

```
kubectl [command] [type] [name] [flags]
```

其中，command 用来指定要对资源执行的操作，例如 create、get 以及 delete 等。type 用来指定资源类型，资源类型是区分字母大小写的。用户可以以单数、复数以及缩略的形式指定资源类型，例如 pod、pods 或者 po 等。name 用来指定资源的名称，资源名称也是区分字母大小写的。如果没有指定资源名称，则默认显示所有的资源。flag 指定可选的参数。例如，可以使用-s 或者-server 参数，指定 Kubernetes API server 的地址和端口。

此外，用户可以通过 kubectl help 命令查找更多的帮助信息，如下所示：

```
    [root@localhost ~]# kubectl help
    kubectl controls the Kubernetes cluster manager.

    Find more information at https://github.com/kubernetes/kubernetes.

    Basic Commands (Beginner):
      create        Create a resource by filename or stdin
      expose        Take a replication controller, service, deployment or pod and
expose it as a new Kubernetes Service
      run           Run a particular image on the cluster
```

```
      set           Set specific features on objects

    Basic Commands (Intermediate):
      get           Display one or many resources
      explain       Documentation of resources
      edit          Edit a resource on the server
      delete        Delete resources by filenames, stdin, resources and names, or
by resources and label selector

    Deploy Commands:
      rollout       Manage a deployment rollout
      rolling-update Perform a rolling update of the given ReplicationController
      scale         Set a new size for a Deployment, ReplicaSet, Replication
Controller, or Job
      autoscale     Auto-scale a Deployment, ReplicaSet, or
ReplicationController

    Cluster Management Commands:
      certificate   Modify certificate resources.
      cluster-info  Display cluster info
      top           Display Resource (CPU/Memory/Storage) usage
      cordon        Mark node as unschedulable
      uncordon      Mark node as schedulable
      drain         Drain node in preparation for maintenance
    …
```

3.1.2　kubectl 的子命令

kubectl 作为 Kubernetes 的命令行工具，主要的职责就是对集群中的资源对象进行操作，这些操作包括对资源对象的创建、删除和查看等。因此，kubectl 提供了非常多的子命令。表 3-1 显示了 kubectl 支持的所有命令，以及这些命令的语法和描述信息。

表 3-1　kubectl的常用子命令

命　　令	语　　法	说　　明
annotate	kubectl annotate (-f filename | type name | type/name) key_1=val_1 … key_n=val_n [–overwrite] [–all] [–resource-version=version] [flags]	添加或更新一个或多个资源注释
api-versions	kubectl api-versions [flags]	列出可用的 api 版本
apply	kubectl apply -f filename [flags]	将来自于文件或标准输入的配置变更应用到主要对象中
attach	kubectl attach pod -c container [-i] [-t] [flags]	连接到正在运行的容器上，以查看输出流或与容器交互

（续表）

命　　令	语　　法	说　　明
autoscale	kubectl autoscale (-f filename | type name | type/name) [–min=minpods] –max=maxpods [–cpu-percent=cpu] [flags]	自动扩容和缩容由副本控制器管理的 Pod
cluster-info	kubectl cluster-info [flags]	显示群集中的主节点和服务的端点信息
config	kubectl config subcommand [flags]	修改 kubeconfig 文件
create	kubectl create -f filename [flags]	从文件或标准输入中创建一个或多个资源对象
delete	kubectl delete (-f filename | type [name | /name | -l label | –all]) [flags]	删除资源对象
describe	kubectl describe (-f filename | type [name_prefix | /name | -l label]) [flags]	显示一个或者多个资源对象的详细信息
edit	kubectl edit (-f filename | type name | type/name) [flags]	通过默认编辑器编辑和更新服务器上的一个或多个资源对象
exec	kubectl exec pod [-c container] [-i] [-t] [flags] [– command [args…]]	在 Pod 的容器中执行一个命令
explain	kubectl explain [–include-extended-apis=true] [–recursive=false] [flags]	获取 Pod、Node 和服务等资源对象的文档
expose	kubectl expose (-f filename | type name | type/name) [–port=port] [–protocol=tcp|udp] [–target-port=number-or-name] [–name=name] [—-external-ip=external-ip-of-service] [–type=type] [flags]	为副本控制器、服务或 Pod 等暴露一个新的服务
get	kubectl get (-f filename | type [name | /name | -l label]) [–watch] [–sort-by=field] [[-o | –output]=output_format] [flags]	列出一个或多个资源
label	kubectl label (-f filename | type name | type/name) key_1=val_1 … key_n=val_n [–overwrite] [–all] [–resource-version=version] [flags]	添加或更新一个或者多个资源对象的标签
logs	kubectl logs pod [-c container] [–follow] [flags]	显示 Pod 中一个容器的日志
patch	kubectl patch (-f filename | type name | type/name) –patch patch [flags]	使用策略合并补丁来更新资源对象中的一个或多个字段
port-forward	kubectl port-forward pod [local_port:]remote_port […[local_port_n:]remote_port_n] [flags]	将一个或多个本地端口转发到 Pod
proxy	kubectl proxy [–port=port] [–www=static-dir] [–www-prefix=prefix] [–api-prefix=prefix] [flags]	为 kubernetes api 服务器运行一个代理
replace	kubectl replace -f filename	从文件或 stdin 中替换资源对象

（续表）

命　　令	语　　法	说　　明
rolling-update	kubectl rolling-update old_controller_name ([new_controller_name] –image=new_container_image | -f new_controller_spec) [flags]	通过逐步替换指定的副本控制器和 Pod 来执行滚动更新
run	kubectl run name –image=image [–env="key=value"] [–port=port] [–replicas=replicas] [–dry-run=bool] [–overrides=inline-json] [flags]	在集群上运行一个指定的镜像
scale	kubectl scale (-f filename | type name | type/name) –replicas=count [–resource-version=version] [–current-replicas=count] [flags]	扩容和缩容副本集的数量
version	kubectl version [–client] [flags]	显示运行在客户端和服务器端的 kubernetes 版本

3.1.3　Kubernetes 资源对象类型

在 Kubernetes 中，提供了很多的资源对象，开发和运维人员可以通过这些对象对容器进行编排。表 3-2 是 kubectl 所支持的资源对象类型，以及它们的缩略别名。

表 3-2　kubectl所支持的资源对象列表

资　　源	缩　　写	说　　明
clusters		集群
componentstatuses	cs	组件对象状态
configmaps	cm	configmaps 可以被用来保存单个属性，也可以用来保存整个配置文件或者 JSON 二进制对象。ConfigMap API 资源存储“键-值对”配置数据，这些数据可以在 pods 里使用
daemonsets	ds	daemonsets 能够让所有（或者一些特定）的 Node 节点运行同一个 Pod
deployments	deploy	deployments 是 Kubernetes 中的一种控制器，是比 ReplicaSet 更高级的概念，它最重的特性是支持对 Pod 与 ReplicaSet 的声明式升级，声明式升级比其他方式的升级更安全可靠
endpoints	ep	endpoints 是实现实际服务的端点集合
events	ev	记录了集群运行所产生的各种事件
ingress	ing	ingress 是 k8s 集群中的一个 API 资源对象，扮演边缘路由器（edge router）的角色
nodes	no	节点
namespaces	ns	命名空间
pods	po	获取 Pod 信息
replicasets	rs	代用户创建指定数量的 Pod 副本数量，确保 Pod 副本数量符合预期状态，并且支持滚动式自动扩容和缩容功能
cronjob		周期性任务控制，不需要持续后台运行
services	svc	各种服务

3.1.4 kubectl 输出格式

默认情况下，kubectl 命令输出格式为纯文本。但是，用户可以通过-o 或者--output 选项来指定其他的输出格式。表 3-3 列出了常见的输出格式及其选项名。

表 3-3 kubectl的输出格式

选 项	说 明
custom-columns=<spec>	根据自定义列名进行输出，用逗号分隔
custom-columns-file=<filename>	从文件中获取自定义列名进行输出
json	以 JSON 格式显示结果
jsonpath=<template>	输出 jasonpath 表达式定义的字段信息
jasonpath-file=<filename>	输出 jsonpath 表达式定义的字段信息，来源于文件
name	仅输出资源对象的名称
wide	输出更多信息，比如会输出节点名
yaml	以 yaml 格式输出

3.1.5 kubectl 命令举例

为了使读者能够快速掌握 kubectl 命令的使用方法，下面对常用的命令进行介绍。

1. kubectl create 命令

此命令通过文件或者标准输入创建一个资源对象，支持 YAML 或者 JSON 格式的配置文件。例如，如果用户创建了一个 Nginx 的 YAML 配置文件，其内容如下：

```
apiVersion: v1
kind: ReplicationController
metadata:
  name: nginx-controller
spec:
  replicas: 2
  selector:
    name: nginx
  template:
    metadata:
      labels:
        name: nginx
    spec:
      containers:
        - name: nginx
          image: nginx
          ports:
            - containerPort: 80
```

用户可以使用以下命令创建 MySQL 的副本控制器：

```
[root@localhost ~]# kubectl create -f nginx.yaml
replicationcontroller "nginx-controller" created
```

2．kubectl get 命令

用户可以通过此命令列出一个或多个资源对象，该命令的参数为资源类型名称。例如，下面的命令列出当前命名空间下的节点：

```
[root@localhost ~]# kubectl get nodes
NAME              STATUS     AGE
192.168.1.122        Ready        12d
192.168.1.123        Ready        12d
```

下面的命令列出所有的服务：

```
[root@localhost ~]# kubectl get services
NAME            CLUSTER-IP      EXTERNAL-IP         PORT(S)     AGE
kubernetes      10.254.0.1          <none>          443/TCP     12d
…
```

下面的命令以比较详细的方式列出当前命名空间中的 Pod：

```
[root@localhost ~]# kubectl get pods -o wide
NAME                READY   STATUS  RESTARTS     AGE      IP            NODE
nginx-controller-g5165      1/1       Running  0          4m  172.17.0.3
127.0.0.1
nginx-controller-zz0dk      1/1       Running  0          4m      172.17.0.2
127.0.0.1
…
```

3．kubectl describe 命令

此命令用于显示一个或多个资源对象的详细信息。例如，我们想要获取名为 nginx-controller-g5165 的 Pod 的详细信息，可以使用以下命令：

```
[root@localhost ~]# kubectl describe pods/nginx-controller-g5165
Name:           nginx-controller-g5165
Namespace:       default
Node:           127.0.0.1/127.0.0.1
Start Time:      Sat, 23 Mar 2019 07:00:19 +0800
Labels:          name=nginx
Status:          Running
IP:             172.17.0.3
Controllers:     ReplicationController/nginx-controller
Containers:
  nginx:
```

```
    Container ID:
docker://5c427e437a4cb413e33b12c3dcc5d16ec1772876a4e5552669d842edf8bb1372
    Image:                      nginx
    Image ID:                   docker-pullable://docker.io/nginx@sha256:
98efe605f61725fd817ea69521b0eeb32bef007af0e3d0aeb6258c6e6fe7fc1a
    Port:                       80/TCP
    State:                      Running
      Started:                  Sat, 23 Mar 2019 07:03:37 +0800
    Ready:                      True
    Restart Count:              0
    Volume Mounts:              <none>
    Environment Variables:      <none>
Conditions:
  Type          Status
  Initialized   True
  Ready         True
  PodScheduled  True
No volumes.
…
```

4．kubectl exec 命令

此命令用于在 Pod 中的容器上执行一个命令。例如，下面的命令在名为 my-nginx-379829228-8gfbb 的容器上执行/bin/bash 命令：

```
[root@localhost ~]# kubectl exec -it my-nginx-379829228-8gfbb /bin/bash
root@my-nginx-379829228-8gfbb:/#
$ kubectl exec -it nginx-c5cff9dcc-dr88w /bin/bash
```

执行完以后，可以发现 Shell 的命令提示符发生了变化，表明已经进入了容器的 Shell 环境中。

如果想要在容器中执行 ls 命令，则可以使用以下方式：

```
[root@localhost ~]# kubectl exec nginx-controller-g5165 ls
bin
boot
dev
etc
home
lib
lib64
media
…
```

在上面的命令中，nginx-controller-g5165 为 Pod 的名称。

5. kubectl run 命令

该命令用来创建一个应用。与 kubectl create 命令不同，在该命令中，所有的选项可以通过命令行指定。例如，下面的命令创建一个 Nginx 应用：

```
[root@localhost ~]# kubectl run --image=nginx nginx-app --port=8080
deployment "nginx-app" created
```

执行完之后，通过 get 命令查看创建进度，如下所示：

```
[root@localhost ~]# kubectl get pods
NAME                        READY       STATUS      RESTARTS        AGE
nginx-app-2743647498-f6qc6 1/1          Running     0                  1m
…
```

从上面的输出结果可知，刚刚创建的 Pod 已经处于运行状态。

6. kubectl delete 命令

该命令用来删除集群中的资源。例如，下面的命令删除名为 nginx-controller-zz0dk 的 Pod：

```
[root@localhost ~]# kubectl delete pods/nginx-controller-zz0dk
pod "nginx-controller-zz0dk" deleted
```

除了上面介绍的几个命令之外，kubectl 还提供了许多功能强大的命令，读者可以参考其他的技术文档，在此不再详细介绍。

3.2　kubeadm 的使用方法

kubeadm 是 Kubernetes 官方主推的用于快速安装 Kubernetes 集群的命令行工具。伴随 Kubernetes 每个版本的发布都会同步更新，kubeadm 会对集群配置方面的一些实践进行调整。本节将详细介绍 kubeadm 的使用方法。

3.2.1　kubeadm 安装方法

默认情况下，kubeadm 不会被自动安装。如果用户想要使用 kubeadm 工具，则可以下载 Kubernetes 的二进制文件包，当然也可以通过源代码自己编译生成。如果已经下载了 Kubernetes 的二进制文件，则 kubeadm 命令位于 server/bin 目录中。从下面的列表可以发现，该目录中包含了多个 Kubernetes 的可执行文件，例如 kube-apiserver、kubectl 以及 kubelet 等。

```
[root@localhost bin]# ll
total 1405788
-rwxr-xr-x  1   root        root    39847392    Feb 28 22:12
apiextensions-apiserver
…
-rwxr-xr-x  1   root        root        177790872   Feb 28 22:12
hyperkube
```

```
-rwxr-xr-x  1   root        root        36415584    Feb 28 22:12    kubeadm
-rwxr-xr-x  1   root        root        138661120   Feb 28 22:12
kube-apiserver
-rw-r--r--  1   root        root        182551552   Feb 28 22:08
kube-apiserver.tar
-rwxr-xr-x  1   root        root        103921568   Feb 28 22:12
kube-controller-manager
…
-rwxr-xr-x  1   root        root    39251424    Feb 28 22:12    kubectl
-rwxr-xr-x  1   root        root        113031192   Feb 28 22:12    kubelet
-rwxr-xr-x  1   root        root    34832736    Feb 28 22:12    kube-proxy
-rw-r--r--      1   root        root    82128896    Feb 28 22:08
kube-proxy.tar
-rwxr-xr-x  1   root        root        37300480    Feb 28 22:12
kube-scheduler
-rw-r--r--      1   root        root    81190912    Feb 28 22:08
kube-scheduler.tar
-rwxr-xr-x  1   root        root    1595200         Feb 28 22:12    mounter
```

除此之外，如果是在 CentOS 中安装 kubeadm，用户还可以通过软件包管理工具来安装。

先添加 Kubernetes 的软件仓库（或称为软件存储库），创建/etc/yum.repos.d/kubernetes.repo 文件，其内容如下：

```
[kubernetes]
name=Kubernetes
baseurl=https://mirrors.aliyun.com/kubernetes/yum/repos/kubernetes-el7-x8
6_64
enabled=1
gpgcheck=1
repo_gpgcheck=1
gpgkey=https://mirrors.aliyun.com/kubernetes/yum/doc/yum-key.gpg
https://mirrors.aliyun.com/kubernetes/yum/doc/rpm-package-key.gpg
```

上面的代码将阿里云的 Kubernetes 镜像站点添加到当前系统中。然后通过以下命令安装：

```
[root@localhost ~]# yum install -y kubeadm
```

3.2.2 kubeadm 基本语法

kubeadmin 的基本语法如下：

```
kubeadm [command]
```

其中 command 为 kubeadm 提供的子命令，常用的子命令有：

- config：指定初始化集群时使用的配置文件。
- init：初始化 Master 节点。

- join：初始化 Node 节点并加入集群。
- reset：重置当前节点，包括 Master 节点和 Node 节点。

3.2.3　部署 Master 节点

在使用 kubeadm 命令之前，用户首先需要处理几个先决条件，这是因为 kubeadm 依赖于 Docker 和 kubelet 等组件服务，下面分别进行介绍。

1．安装 Docker

用户需要在节点上面安装 Docker，命令如下：

```
[root@localhost ~]# yum -y install docker
```

然后通过以下命令启动 Docker：

```
[root@localhost ~]# systemctl enable docker
[root@localhost ~]# systemctl start docker
```

2．安装 kubelet

接下来使用以下命令安装 kubelet 以及其他组件：

```
[root@localhost ~]# yum install -y kubelet kubeadm kubectl ipvsadm
```

在节点上启动 kubelet 服务：

```
[root@localhost ~]# systemctl enable kubelet
[root@localhost ~]# systemctl start kubelet
```

3．禁用 SELinux

用户还需要禁用 SELinux，命令如下：

```
[root@localhost ~]# sed -i 's/SELINUX=permissive/SELINUX=disabled/'
/etc/sysconfig/selinux
[root@localhost ~]# setenforce 0
setenforce: SELinux is disabled
```

4．禁用交换分区

用户可以通过以下命令临时禁用交换分区：

```
[root@localhost ~]# swapoff -a
```

然后修改/etc/fstab 文件，将其中关于交换分区的项目注释掉，防止操作系统重新启动后自动挂载交换分区，如下所示：

```
#/dev/mapper/centos-swap swap                    swap    defaults        0 0
```

5．修改防火墙规则

Docker 从 1.13 版本开始调整了默认的防火墙规则，禁用了 iptables 的 filter 表中 FOWARD

链，这样会引起 Kubernetes 集群中跨节点的 Pod 无法通信，所以用户需要修改该规则，命令如下：

```
[root@localhost ~]# iptables -P FORWARD ACCEPT
[root@localhost ~]# iptables-save
```

6. 配置转发

创建/etc/sysctl.d/k8s.conf 文件，其内容如下：

```
net.bridge.bridge-nf-call-ip6tables = 1
net.bridge.bridge-nf-call-iptables = 1
vm.swappiness=0
```

调用以下命令使配置生效：

```
[root@localhost ~]# sysctl --system
```

7. 修改 kubelet 配置文件

编辑/etc/sysconfig/kubelet 文件，在 KUBELET_EXTRA_ARGS 配置项中增加以下代码：

```
--cgroup-driver=systemd
--pod-infra-container-image=registry.cn-hangzhou.aliyuncs.com/google_containers/pause-amd64:3.1
```

上述代码的作用是使 Kubernetes 中的 Pause 容器使用国内的镜像。

重新启动 kubelet：

```
[root@localhost ~]# systemctl daemon-reload
[root@localhost ~]# systemctl enable kubelet && systemctl restart kubelet
```

注　意

上面的操作需要在所有的节点上执行。

处理好所有的先决条件之后，就可以部署 Master 节点了，命令如下：

```
[root@localhost ~]# kubeadm init
[init] Using Kubernetes version: v1.14.1
[preflight] Running pre-flight checks
[preflight] Pulling images required for setting up a Kubernetes cluster
[preflight] This might take a minute or two, depending on the speed of your internet connection
[preflight] You can also perform this action in beforehand using 'kubeadm config images pull'
```

接下来就是等待过程，在这个过程中，kubeadm 会完成以下几个主要步骤：

（1）检查初始化节点所需要的先决条件，如果不满足，则给出错误提示。

（2）生成 Kubernetes 集群的令牌。

（3）生成自签名的 CA 和客户端证书。

（4）自动创建 kubeconfig 配置文件，该文件是提供给 kubelet 连接 API Server 时使用的。

（5）配置基于角色的访问控制，并且设置 Maser 节点只允许控制组件服务。

（6）创建其他的相关服务，例如 kube-proxy 和 kube-dns 等。

但是 kubeadm 命令并不初始化网络组件，所以对于网络方面的插件，需要用户单独配置。

3.2.4　部署 Node 节点

部署普通的 Node 节点需要使用 join 命令，在使用该命令时，需要提供前面初始化 Master 节点时生成的令牌。命令如下：

```
[root@localhost ~]# token=$(kubeadm token list | grep authentication,signing | awk '{print $1}')
[root@localhost ~]# kubeadm join --token $token 192.168.21.138
```

在上面的命令中，第 1 行用来获取令牌，第 2 行执行初始化并加入集群，--token 选项用来指定令牌，后面的 IP 地址为 Master 节点的 IP 地址。

在上面的过程中，kubeadm 命令会自动从 API 服务器上面下载证书，然后创建本地证书，请求签名，最后配置 kubelet 服务注册到 API 服务器。

3.2.5　重置节点

无论是 Master 节点还是 Node 节点，如果想要重新部署，都可以使用 reset 命令，如下所示：

```
[root@localhost ~]# kubeadm reset
[reset] WARNING: Changes made to this host by 'kubeadm init' or 'kubeadm join' will be reverted.
[reset] Are you sure you want to proceed? [y/N]: y
[preflight] Running pre-flight checks
W0428 02:26:34.961358   38084 reset.go:234] [reset] No kubeadm config, using etcd pod spec to get data directory
[reset] No etcd config found. Assuming external etcd
[reset] Please manually reset etcd to prevent further issues
[reset] Stopping the kubelet service
[reset] unmounting mounted directories in "/var/lib/kubelet"
[reset] Deleting contents of stateful directories: [/var/lib/kubelet /etc/cni/net.d /var/lib/dockershim /var/run/kubernetes]
[reset] Deleting contents of config directories: [/etc/kubernetes/manifests /etc/kubernetes/pki]
[reset] Deleting files: [/etc/kubernetes/admin.conf /etc/kubernetes/kubelet.conf /etc/kubernetes/bootstrap-kubelet.conf /etc/kubernetes/controller-manager.conf /etc/kubernetes/scheduler.conf]
```

第4章 运行应用

通过前面几章的学习，读者应该对 Kubernetes 有了充分的认识。Kubernetes 作为一种容器编排引擎，最重要的功能就是管理好容器，提供各种服务。在本章中，我们将详细介绍如何在 Kubernetes 中部署各种容器化应用。

本章涉及的知识点主要有：

- Deployment 及其使用方法：主要介绍如何通过创建和运行 Deployment，以及如何用 Deployment 管理 ReplicaSet、Pod，实现滚动升级、回滚应用、扩容和缩容。
- DaemonSet 的使用方法：主要介绍 DaemonSet 的基本概念、Kubernetes 系统中的 DaemonSet 以及如何运行自己的 DaemonSet。
- Job：主要介绍 Job 的用途，Job 的并行性以及定时 Job 等。

4.1 Deployment

Deployment 提供了一种更加简单的、更新 Replication Controller 和 Pod 的机制，更好地解决了 Pod 的编排问题。本节将详细介绍如何通过 Deployment 实现 Pod 的管理。

4.1.1 什么是 Deployment

Deployment 的中文意思为部署、调度，它是在 Kubernetes 的版本 1.2 中新增加的一个核心概念。Deployment 的实现为用户管理 Pod 提供了一种更为便捷的方式。用户可以通过在 Deployment 中描述所期望的集群状态，Deployment 会将现在的集群状态在一个可控的速度下逐步更新成所期望的集群状态。与 Replication Controller 基本一样，Deployment 主要职责同样是为了保证 Pod 的数量和健康。Deployment 的绝大部分的功能与 Replication Controller 完全一样，因此，用户可以将 Deployment 看作是升级版的 Replication Controller。

与 Replication Controller 相比，除了继承 Replication Controller 的全部功能之外，Deployment 还有以下新的特性：

- 事件和状态查看：可以查看 Deployment 的升级详细进度和状态。
- 回滚：当升级 Pod 镜像或者相关参数的时候发现问题，可以使用回滚操作回滚到上一个稳定的版本或者指定的版本。

- 版本记录：每一次对 Deployment 的操作，都能保存下来，给予后续可能的回滚使用。
- 暂停和启动：对于每一次升级，都能够随时暂停和启动。
- 多种升级方案：主要包括重建，即删除所有已存在的 Pod，重新创建新的 Pod；滚动升级，即采用逐步替换的策略来升级 Pod，Pod 的滚动升级方案可以支持更多的附加参数。

4.1.2 Deployment 与 ReplicaSet

说到 ReplicaSet 对象，还是不得不提 Replication Controller。在 Kubernetes v1.16 之前，只有 Replication Controller 对象，它的主要作用是确保 Pod 以用户指定的副本数运行，即如果有容器异常退出，Replication Controller 会自动创建新的 Pod 来替代，而异常多出来的容器也会自动回收。可以说，通过 Replication Controller，Kubernetes 实现了集群的高可用性。

在新版本的 Kubernetes 中，建议使用 ReplicaSet 来取代 Replication Controller。ReplicaSet 跟 Replication Controller 没有本质的不同，只是名字不一样，并且 ReplicaSet 支持集合式的选择器，而 Replication Controller 只支持等式选择器。

虽然 ReplicaSet 也可以独立使用，但是 Kubernetes 并不建议用户直接操作 ReplicaSet 对象。而是通过更高层次的对象 Deployment 来自动管理 ReplicaSet，这样就无需担心跟其他机制的不兼容问题，比如 ReplicaSet 不支持滚动更新，但 Deployment 支持，并且 Deployment 还支持版本记录、回滚、暂停升级等高级特性。这意味着用户几乎不会有机会去直接管理 ReplicaSet。

注　　意
用户不该手动管理由 Deployment 创建的 ReplicaSet，否则就“篡越”了 Deployment 的职责。

4.1.3 运行 Deployment

我们先从一个最简单的例子开始，介绍如何运行 Deployment，命令如下：

```
[root@localhost ~]# kubectl run nginx-deployment --image=nginx:1.7.9
--replicas=3
deployment "nginx-deployment" created
```

在上面的命令中，nginx-deployment 是要创建的 Deployment 的名称，--image 选项用来指定容器所使用的镜像，其中 nginx 为镜像名称，1.7.9 为版本号。replicas 选项用来指定 Pod 的副本数为 3，即当前集群中在任何时候都要保证有 3 个 Nginx 的副本在运行。

执行完命令之后，用户可以通过 get 命令查看刚才运行的 Deployment，如下所示：

```
[root@localhost ~]# kubectl get deployment nginx-deployment
NAME              DESIRED    CURRENT    UP-TO-DATE    AVAILABLE   AGE
nginx-deployment   3         3          3             3        10s
```

在上面的输出结果中，DESIRED 表示预期的副本数，即我们在命令中通过--replicas 参数指定的数量。CURRENT 表示当前的副本数。实际上 CURRENT 是 Deployment 所创建的 ReplicaSet 里面的 Replica 的值，在部署的过程中，这个数值会不断增加，一直增加到 DESIRED

所指定的值为止。UP-TO-DATE 表示当前已经处于最新版本的 Pod 的副本数，主要用于在滚动升级的过程中，表示当前已经有多少个副本已经成功升级。AVAILABLE 表示当前集群中属于当前 Deployment 的、可用的 Pod 的副本数量，实际上就是当前 Deployment 所产生的，在当前集群中存活的 Pod 的数量。AGE 表示当前 Deployment 的年龄，即从创建到现在的时间差。

接下来我们通过 describe deployment 命令查看当前 Deployment 更加详细的信息，如下所示：

```
[root@localhost ~]# kubectl describe deployment nginx-deployment
Name:                   nginx-deployment
Namespace:              default
CreationTimestamp:      Sun, 24 Mar 2019 06:57:18 +0800
Labels:                 run=nginx-deployment
Selector:                   run=nginx-deployment
Replicas:                   3 updated | 3 total | 3 available | 0 unavailable
StrategyType:           RollingUpdate
MinReadySeconds:        0
RollingUpdateStrategy:          1 max unavailable, 1 max surge
Conditions:
  Type          Status      Reason
  ----          ------      ------
  Available     True        MinimumReplicasAvailable
OldReplicaSets: <none>
NewReplicaSet:  nginx-deployment-3954615459 (3/3 replicas created)
No events.
```

上面的输出信息的含义大部分都非常明确，我们重点关注几点即可。首先 Namespace 表示当前 Deployment 所属的命名空间为 default，即缺省的命名空间。Replicas 表示当前 Pod 的副本信息。NewReplicaSet 表示由 Deployment 自动创建的 ReplicaSet，其名称为 nginx-deployment-3954615459，可以发现实际上 ReplicaSet 的名称使用所属的 Deployment 的名称作为前缀，这样可以非常容易地分辨这个 ReplicaSet 是由哪个 Deployment 创建的。

前面已经介绍过的 ReplicaSet 是一个非常重要的概念，而且用户几乎不会直接操作 ReplicaSet，可以通过 Deployment 间接地管理它。下面我们通过 get replicaset 命令来查看 ReplicaSet 的信息，如下所示：

```
[root@localhost ~]# kubectl get replicaset
NAME                          DESIRED     CURRENT     READY      AGE
nginx-deployment-3954615459    3           3           3         10s
```

在上面的输出信息中，ReplicaSet 的名为 nginx-deployment-3954615459，READY 表示当前已经就绪的副本数。

为了了解更多的关于 nginx-deployment-3954615459 的信息，可以使用 describe replicaset 命令，如下所示：

```
[root@localhost ~]# kubectld describe replicaset nginx-deployment-3954615459
Name:           nginx-deployment-3954615459
Namespace:      default
Image(s):       nginx:1.7.9
Selector:       pod-template-hash=3954615459,run=nginx-deployment
Labels:         pod-template-hash=3954615459
                run=nginx-deployment
Replicas:       3 current / 3 desired
Pods Status:    3 Running / 0 Waiting / 0 Succeeded / 0 Failed
No volumes.
No events.
```

Image 表示容器所使用的镜像为 nginx:1.7.9，Replicas 表示副本数，Pods Status 表示当前 Deployment 所创建的 Pod 的状态，可以看到有 3 个副本处于运行（Running）状态，没有处于等待（Waiting）、结束（Succeeded）或者失败（Failed）的 Pod。

接下来，我们继续探讨 Deployment 所创建的 Pod。执行 get pod 命令，获取当前集群中的 Pod 信息，如下所示：

```
[root@localhost ~]# kubectl get pod -o wide
NAME                              READY  STATUS   RESTARTS AGE   IP      NODE
…
nginx-deployment-3954615459-4zp22 1/1     Running     0        1h
172.17.0.5 192.168.1.122
nginx-deployment-3954615459-b2jvj  1/1     Running     0        1h
172.17.0.3 127.0.0.1
nginx-deployment-3954615459-rpjcp  1/1     Running     0        1h
172.17.0.4 127.0.0.1
…
```

从上面的输出结果可知，3 个 Pod 副本都处于运行状态。另外，Kubernetes 还为每个 Pod 自动分配了 IP 地址。其中 1 个副本运行在名称为 192.168.1.122 的节点上面，2 个副本运行在名称为 127.0.0.1 的节点上。为了便于维护，Pod 的命名规则也是以 Deployment 的名称以及 ReplicaSet 的名称作为前缀的。由此可以推断出，在同一个节点上面，可以同时运行一个 Pod 的多个副本。

前面已经介绍过，Kubernetes 中的各种服务最终是由 Pod 里面的容器提供的。因此，了解 Pod 及其容器是我们的最终目标。所以，下面通过 kubectl describe pod 命令查看其中某个 Pod 副本的详细信息，如下所示：

```
[root@localhost ~]# kubectl describe pod nginx-deployment-3954615459-4zp22
Name:           nginx-deployment-3954615459-4zp22
Namespace:      default
Node:           192.168.1.122/192.168.1.122
Start Time:     Sun, 24 Mar 2019 06:57:19 +0800
Labels:         pod-template-hash=3954615459
```

```
                  run=nginx-deployment
    Status:       Running
    IP:           172.17.0.2
    Controllers:  ReplicaSet/nginx-deployment-3954615459
    Containers:
      nginx-deployment:
        Container ID:
docker://df381a4ea070522af06d897d0c49a253def23ac7264a8b5bc3b50fbe67e86b6a
        Image:             nginx:1.7.9
        Image ID:
docker-pullable://docker.io/nginx@sha256:e3456c851a152494c3e4ff5fcc26f240206a
bac0c9d794affb40e0714846c451
        Port:
        State:             Running
          Started:          Sun, 24 Mar 2019 17:02:43 +0800
        Last State:          Terminated
          Reason:           Completed
          Exit Code:         0
          Started:          Sun, 24 Mar 2019 07:59:14 +0800
          Finished:         Sun, 24 Mar 2019 08:03:36 +0800
        Ready:             True
        Restart Count:      1
        Volume Mounts:      <none>
        Environment Variables:   <none>
    Conditions:
      Type        Status
      Initialized  True
      Ready        True
      PodScheduled  True
    No volumes.
    QoS Class:     BestEffort
    Tolerations:   <none>
    Events:
      FirstSeen    LastSeen      Count   From                   SubObjectPath
Type          Reason                Message
      ---------    --------      -----   ----                   -------------
--------      ------                -------
      56m          56m           2       {kubelet 192.168.1.122}
Warning       MissingClusterDNS     kubelet does not have ClusterDNS IP
configured and cannot create Pod using "ClusterFirst" policy. Falling back to
DNSDefault policy.
      56m          56m           1       {kubelet 192.168.1.122}
spec.containers{nginx-deployment}     Normal         Pulled
```

```
Container image "nginx:1.7.9" already present on machine
     56m          56m              1        {kubelet 192.168.1.122}
spec.containers{nginx-deployment}       Normal          Created
Created container with docker id df381a4ea070; Security:[seccomp=unconfined]
     56m          56m              1        {kubelet 192.168.1.122}
spec.containers{nginx-deployment}       Normal          Started
Started container with docker id df381a4ea070
```

该命令的输出信息非常多，我们只要关注其中重要的部分即可。Node 表示当前 Pod 所处的节点。IP 是 Kubernetes 为当前 Pod 副本分配的 IP 地址，用户可以通过该 IP 地址与 Pod 通信。Controllers 表示当前 Pod 由哪个 Deployment 和 ReplicaSet 创建，在本例中，Deployment 的名称为 nginx-deployment，ReplicaSet 的名称为 nginx-deployment-3954615459。Containers 表示当前 Pod 中的容器的列表，本例中只有一个用户容器，其容器 ID 为 docker://df381a4ea070522af06d897d0c49a253def23ac7264a8b5bc3b50fbe67e86b6a。Events 为当前 Pod 副本的日志信息，便于用户调试。

最后，我们的关注点落在了容器上。由于 nginx-deployment-3954615459-4zp22 当前运行在名为 192.168.1.122 的节点上，因此我们登录到 192.168.1.122。然后使用前面介绍的容器管理命令 docker ps，查看当前节点的容器列表，如下所示：

```
    [root@localhost ~]# docker ps
    CONTAINER ID        IMAGE
COMMAND                  CREATED                 STATUS                  PORTS
NAMES
    …
    df381a4ea070        nginx:1.7.9
"nginx -g 'daemon ..."   About an hour ago   Up About an hour
k8s_nginx-deployment.a8dee3c8_nginx-deployment-3954615459-4zp22_default_0271d
1ca-4dbf-11e9-867a-000c2994a2b7_20f4bc78
    fb16d8356f3c
registry.access.redhat.com/rhel7/pod-infrastructure:latest   "/usr/bin/pod"
About an hour ago   Up About an hour
k8s_POD.ae8ee9ac_nginx-deployment-3954615459-4zp22_default_0271d1ca-4dbf-11e9
-867a-000c2994a2b7_5927077c
    …
```

可以看到，在 192.168.1.122 节点上面有 2 个容器，这 2 个容器都是由 nginx-deployment 创建的。还记得我们前面介绍的 Pod 的结构吗？在每个 Pod 中，都存在着一个名为 Pause 的根容器，这个根容器是 Kubernetes 系统的一部分。在上面的容器列表中，ID 为 df381a4ea070 的是用户容器，而 ID 为 fb16d8356f3c 即为当前 Pod 的根容器。

用户可以通过 docker inspect 命令查看容器的详细信息，如下所示：

```
    [root@localhost ~]# docker inspect df381a4ea070
    [
```

```
        {
            "Id":
"df381a4ea070522af06d897d0c49a253def23ac7264a8b5bc3b50fbe67e86b6a",
            "Created": "2019-03-24T09:02:43.579970795Z",
            "Path": "nginx",
            "Args": [
                "-g",
                "daemon off;"
            ],
            "State": {
                "Status": "running",
                "Running": true,
                "Paused": false,
                "Restarting": false,
                "OOMKilled": false,
                "Dead": false,
                "Pid": 13830,
                "ExitCode": 0,
                "Error": "",
                "StartedAt": "2019-03-24T09:02:43.808003Z",
                "FinishedAt": "0001-01-01T00:00:00Z"
            },
            "Image":
"sha256:84581e99d807a703c9c03bd1a31cd9621815155ac72a7365fd02311264512656",
            "ResolvConfPath": "/var/lib/docker/containers/
fb16d8356f3c7e2f630e9f7349791369cbbbfd813a77072e6ac1c3780f2ac710/resolv.conf",
            "HostnamePath": "/var/lib/docker/containers/
fb16d8356f3c7e2f630e9f7349791369cbbbfd813a77072e6ac1c3780f2ac710/hostname",
            "HostsPath": "/var/lib/kubelet/pods/0271d1ca-4dbf-11e9-867a-
000c2994a2b7/etc-hosts",
            "LogPath": "",
            "Name": "/k8s_nginx-deployment.a8dee3c8_nginx-deployment-
3954615459-4zp22_default_0271d1ca-4dbf-11e9-867a-000c2994a2b7_20f4bc78",
            "RestartCount": 0,
            "Driver": "overlay2",
            "MountLabel": "",
            "ProcessLabel": "",
            "AppArmorProfile": "",
            "ExecIDs": null,
            "HostConfig": {
                "Binds": [
                    …
```

由于这部分内容在前面已经详细介绍过，这里不再重复赘述。

至此，我们已经成功地运行了一个 Deployment。总结前面的内容，我们可以发现创建 Deployment 的大致流程。首先，用户通过 kubectl run deployment 命令创建一个 Deployment。接下来，Deployment 会自动根据用户指定的选项，主要是 image 以及 replicas 等创建 ReplicaSet。最后，由 ReplicaSet 创建 Pod 副本和容器。所以，整个流程如图 4-1 所示。

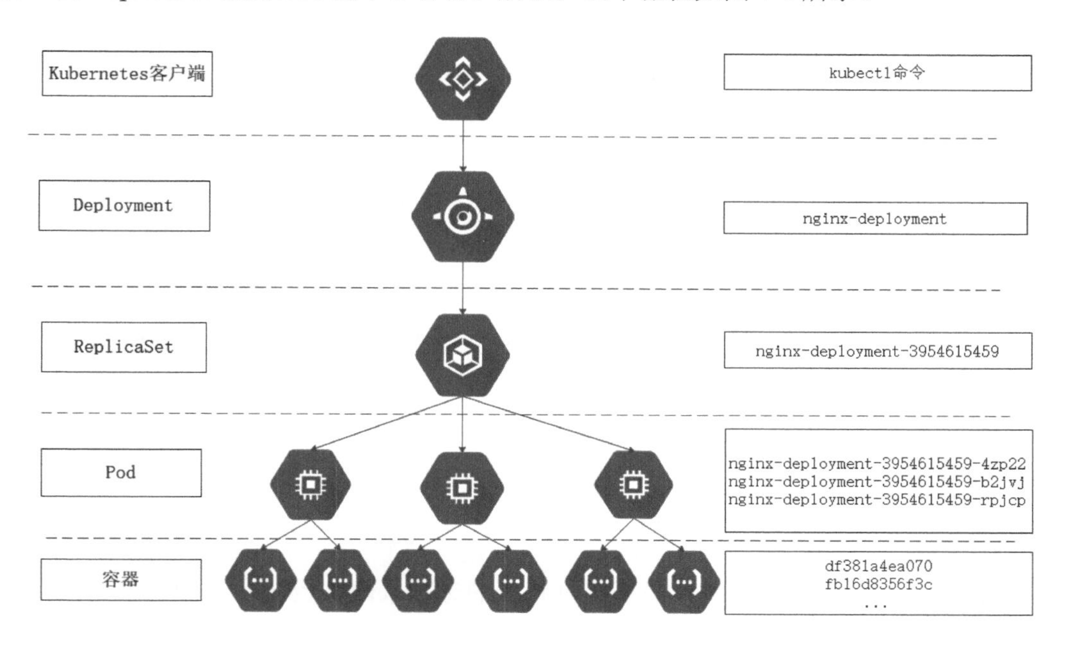

图 4-1　Deployment 创建过程

读者可能会问，我们刚才创建的应用能够提供服务吗？下面我们通过简单的方法来进行验证。刚才我们部署的是一套 Nginx 应用。Nginx 是用来提供各种网络服务的，例如 HTTP 网页服务、反向代理或者网络加速等。这里，我们可以通过 curl 命令来简单测试一下 Nginx 是否运行正常。默认情况下 Nginx 的服务端口为 80，提供 HTTP 访问。我们在任意节点上面执行以下命令：

```
[root@localhost ~]# curl http://172.17.0.3:80
<!DOCTYPE html>
<html>
<head>
<title>Welcome to nginx!</title>
<style>
   body {
      width: 35em;
      margin: 0 auto;
      font-family: Tahoma, Verdana, Arial, sans-serif;
   }
</style>
</head>
```

```
<body>
<h1>Welcome to nginx!</h1>
<p>If you see this page, the nginx web server is successfully installed and
working. Further configuration is required.</p>

<p>For online documentation and support please refer to
<a href="http://nginx.org/">nginx.org</a>.<br/>
Commercial support is available at
<a href="http://nginx.com/">nginx.com</a>.</p>

<p><em>Thank you for using nginx.</em></p>
</body>
</html>
```

curl 命令是一个功能强大的文件传输工具，它支持文件的上传和下载。172.17.0.3 为节点 192.168.1.122 上 Pod 的 IP 地址。可以发现，Nginx 已经可以返回默认的欢迎页面的内容。这表示我们的部署是成功的。

4.1.4 使用配置文件

在前面的例子中，我们直接通过 kubectl 命令创建了一个 Nginx 的 Deployment。可以发现，在使用 kubectl 命令创建资源时，需要提供较多的选项来限制资源的各种属性。实际上，Kubernetes 支持两种方式来定义资源属性，其中一种就是直接在命令行中通过对应的选项来指定，例如，前面的--image=nginx:1.7.3 和--relicas=3 分别用来指定镜像文件和副本数。还有一种方式就是通过配置文件。Kubernetes 支持多种类型的配置文件，常见的有 YAML 和 JSON。

YAML 是一个可读性高，用来表达数据和信息序列的编程语言。YAML 参考了其他多种语言，包括 XML、C 语言、Python、Perl 以及电子邮件格式 RFC2822。YAML 强调这种语言以数据为中心，而不是以标记语言为重点。

YAML 文件的后缀可以使用.yml，也可以使用.yaml，官方推荐使用.yaml。Kubernetes 里所有的资源或者配置文件都可以用 YAML 或 JSON 定义。YAML 是 JSON 的超集，任何有效的 JSON 文件也都是一个有效的 YAML 文件。

YAML 文件具有以下语法规范：

- 区分字母大小写。
- 使用缩进表示层级关系。
- 缩进时不允许使用制表符，只允许使用空格。
- 缩进的空格数目不重要，只要相同层级的元素左侧对齐即可。
- #表示注释，从这个字符一直到行尾，都会被解析器忽略。

YAML 有两种比较重要的数据结构，分别为 Map 和 List，用户只要掌握好这两种数据结构就可以了。

首先介绍一下 Map。跟其他的程序设计语言一样，Map 类型的数据结构通常就是一个“键-值对”（Key-Value Pair）。用户可以非常方便地通过键去引用和修改其对应的值。例如，下面的代码为一个 YAML 的部分内容：

```
apiVersion: v1
kind: Pod
```

在上面的代码中，apiVersion 为键，v1 为其对应的值，表示 Kubernetes 的 API 的版本。kind 同样为键，Pod 为值，表示资源类型为 Pod。当然，用户可以将其转换为具有相同含义的 JSON 格式，如下所示：

```
{
  "apiVersion": "v1",
  "kind": "Pod"
}
```

当然，Map 中的值可以是复杂类型。例如，可以是另外一个 Map 类型的数据结构，如下所示：

```
apiVersion: v1
kind: Pod
metadata:
  name: rss-site
  labels:
    app: web
```

在上面的代码中，键 metadata 的值为一个拥有 2 个键的 Map。上面的代码可以转换为以下 JSON 代码：

```
{
  "apiVersion": "v1",
  "kind": "Pod",
  "metadata": {
             "name": "rss-site",
             "labels": {
                        "app": "web"
                       }
             }
}
```

除了 Map 之外，在表示资源属性时还需要用到一些并列的数据结构，这种情况下，需要使用 YAML 的 List。例如，下面的代码就是一个典型的 List 结构：

```
args
  - sleep
  - "1000"
```

```
  - message
  - "Bring back Firefly!"
```

正如你看到的那样，List 中的列表项的定义以破折号开头，并且与父元素之间存在缩进关系。在 JSON 格式中，它将表示如下：

```
{
   "args": ["sleep", "1000", "message", "Bring back Firefly!"]
}
```

当然，Map 的键值可以是 List 结构，List 的列表项也可以是 Map 结构，如下所示：

```
apiVersion: v1
kind: Pod
metadata:
  name: rss-site
  labels:
    app: web
spec:
  containers:
    - name: front-end
      image: nginx
      ports:
        - containerPort: 80
    - name: rss-reader
      image: nickchase/rss-php-nginx:v1
      ports:
        - containerPort: 88
…
```

以上 YAML 的代码转换为 JSON 后的结构如下：

```
{
   "apiVersion": "v1",
   "kind": "Pod",
   "metadata": {
              "name": "rss-site",
              "labels": {
                         "app": "web"
                        }
             },
    "spec": {
       "containers": [{
                     "name": "front-end",
                     "image": "nginx",
                     "ports": [{
```

```
                    "containerPort": "80"
                    }]
                },
                {
                 "name": "rss-reader",
                 "image": "nickchase/rss-php-nginx:v1",
                 "ports": [{
                    "containerPort": "88"
                    }]
                }]
        }
}
```

在了解了 YAML 的基本语法之后，下面介绍一下 Deployment 中 YAML 的常见格式，如下所示：

```
01 apiVersion: extensions/v1beta1
02 kind: Deployment
03 metadata:
04   name: nginx-deployment-new
05 spec:
06   replicas: 3
07   template:
08     metadata:
09       labels:
10         app: nginx
11         track: stable
12     spec:
13       containers:
14         - name: nginx
15           image: nginx:1.7.9
16           ports:
17             - containerPort: 80
```

第 1 行的 apiVersion 表示当前配置文件格式的版本号。第 2 行 kind 指定当前的资源类型为 Deployment。第 3 行 metadata 用来定义 Deployment 本身的属性，其中 name 表示 Deployment 的名称。第 5 行的 spec 用来定义 Deployment 的规格。用户需要正确区分 metadata 和 spec，前者表示 Deployment 本身固有的属性，后者表示 Deployment 内容的各项属性。第 6 行的 replicas 表示 Pod 的副本数。从第 7 行开始定义 Pod 的模板的各项属性，其中第 10 行定义 Pod 的标签；第 12 行开始定义 Pod 中的容器的各项属性，其中 name 为容器名，image 为容器镜像文件，这 2 个属性是必需的。

将以上代码保存为 nginx-deployment.yaml，然后通过以下命令创建 Deployment：

```
[root@localhost ~]# kubectl apply -f nginx-deployment.yaml
deployment "nginx-deployment-new" created
```

执行完成之后，用户就可以查看 nginx-deployment-new 所创建的各种资源，如下所示：

```
[root@localhost ~]# kubectl get deployments
NAME                 DESIRED    CURRENT    UP-TO-DATE  AVAILABLE    AGE
nginx-deployment        3          3          3           3         13h
nginx-deployment-new    3          3          3           3         5m
[root@localhost ~]# kubectl get replicaset
NAME                            DESIRED    CURRENT    READY        AGE
nginx-deployment-3954615459        3          3          3          13h
nginx-deployment-new-94093859      3          3          3          5m
[root@localhost ~]# kubectl get pods
NAME                            READY   STATUS  RESTARTS        AGE
nginx-deployment-3954615459-4zp22       1/1         Running     1      13h
nginx-deployment-3954615459-b2jvj       1/1         Running     1      13h
nginx-deployment-3954615459-rpjcp       1/1         Running     1      13h
nginx-deployment-new-94093859-395sf     1/1         Running     0      5m
nginx-deployment-new-94093859-d6wrx     1/1         Running     0      5m
nginx-deployment-new-94093859-qnzbb     1/1         Running     0      5m
```

可以发现，使用命令行参数和配置文件都可以帮助用户创建各种资源。命令行参数使用起来非常便捷，用户不需要花费时间去编辑 YAML 文件。而使用 YAML 配置文件也有着明显的优势：

（1）配置文件非常详细地描述了我们想要创建的资源以及最终要达到的状态。

（2）配置文件提供了创建资源的模板，可以重复利用。

（3）由于配置文件可以长久地存储在磁盘上面，因此用户可以像管理代码一样管理部署。

（4）配置文件非常适合于正式的、跨环境的以及大规模的部署。

因此，用户应该尽可能地使用配置文件来创建各类资源。

4.1.5 扩容和缩容

扩容和缩容是指在线增加或者减少 Pod 的副本数量。在前面的例子中，我们指定了 3 个 Nginx 副本，如下所示：

```
    [root@localhost ~]# kubectl get pods -o wide
    NAME                            READY  STATUS   RESTARTS  AGE IP     NODE
    nginx-deployment-new-94093859-395sf  1/1   Running   0       1h  172.17.0.5
127.0.0.1
    nginx-deployment-new-94093859-d6wrx 1/1    Running   0       1h 172.17.0.6
192.168.1.122
    nginx-deployment-new-94093859-qnzbb 1/1    Running   0       1h172.17.0.4
192.168.1.122
```

这3个副本中，有1个副本运行在节点127.0.0.1上，2个副本运行在192.168.1.122上。下面我们修改nginx-deployment.yaml配置文件，增加一个副本：

```
01 apiVersion: extensions/v1beta1
02 kind: Deployment
03 metadata:
04   name: nginx-deployment-new
05 spec:
06   replicas: 4
07   template:
08     metadata:
09       labels:
10         app: nginx
11         track: stable
12     spec:
13       containers:
14         - name: nginx
15           image: nginx:1.7.9
16           ports:
17             - containerPort: 80
```

然后执行以下命令更新Deployment：

```
[root@localhost ~]# kubectl apply -f nginx-deployment.yaml
deployment "nginx-deployment-new" configured
[root@localhost ~]# kubectl get deployments
NAME                    DESIRED     CURRENT     UP-TO-DATE  AVAILABLE   AGE
…
nginx-deployment-new    4           4           4           4           33m
```

从上面的命令可知，当前Deployment的预期副本数已经变成4，当前可用的副本数也为4。为了验证详细的Pod副本数量，用户可以使用kubectl describe pods命令，如下所示：

```
[root@localhost ~]# kubectl get pods -o wide
NAME                                     RE  STATUS RE AGE IP          NODE
nginx-deployment-new-94093859-395sf     1/1 Running     0   1h  172.17.0.5
 127.0.0.1
nginx-deployment-new-94093859-72030     1/1 Running     0   9s  172.17.0.7
127.0.0.1
nginx-deployment-new-94093859-d6wrx     1/1 Running     0   1h  172.17.0.6
192.168.1.122
nginx-deployment-new-94093859-qnzbb     1/1 Running     0   1h  172.17.0.4
192.168.1.122
```

从上面的输出结果可知，当前的 Deployment 有 4 个副本，其中 127.0.0.1 和 192.168.1.122 上各有 2 个。这意味着 Kubernetes 根据修改后的配置文件，新增了一个副本，并且将其调度在 127.0.0.1 节点上运行。

Deployment 的扩容也可以通过命令行快速完成，这需要使用 kubectl scale 命令，在该命令中指定需要扩展到的副本数，如下所示：

```
    [root@localhost ~]# kubectl scale deployment nginx-deployment-new
--replicas=5
    deployment "nginx-deployment-new" scaled
```

上面的命令将副本数扩大到 5，执行完成后查看 Pod 信息，如下所示：

```
    [root@localhost ~]# kubectl get pods -o wide
    NAME                             RE  STA      RES AGE IP          NODE
    nginx-deployment-new-94093859-395sf    1/1 Running     0   1h   172.17.0.5
    127.0.0.1
    nginx-deployment-new-94093859-72030    1/1 Running     0   20m 172.17.0.6
    127.0.0.1
    nginx-deployment-new-94093859-d6wrx    1/1 Running     0   1h   172.17.0.5
192.168.1.122
    nginx-deployment-new-94093859-qnzbb    1/1 Running 0    1h   172.17.0.4
192.168.1.122
    nginx-deployment-new-94093859-v3fq4    1/1 Running 0    9s   172.17.0.6
192.168.1.122
```

从上面的输出结果可知，Pod 副本数已经变成了 5 个。

Deployment 的缩容的操作与扩容相反。为了减少副本数量，用户需要修改 nginx-deployment.yaml 配置文件，将其中的 replicas 修改为 3，然后执行 kubectl apply 命令，即可完成缩容操作，如下所示：

```
    [root@localhost ~]# kubectl apply -f nginx-deployment2.yaml
    deployment "nginx-deployment-new" configured
```

执行完成之后，通过 kubectl get pods 命令查看执行结果，如下所示：

```
    [root@localhost ~]# kubectl get pods -o wide
    NAME                             RE  ST       RES AGE IP         NODE
    nginx-deployment-new-94093859-395sf    1/1  Running    0   1h   172.17.0.5
127.0.0.1
    nginx-deployment-new-94093859-d6wrx    1/1 Running     0   1h   172.17.0.5
192.168.1.122
    nginx-deployment-new-94093859-qnzbb 1/1    Running     0   1h   172.17.0.4
192.168.1.122
```

可以看到有 2 个副本已经被删除，只保留了 3 个副本。

同样的操作也可以使用命令行完成，如下所示：

```
[root@localhost ~]# kubectl scale deployment nginx-deployment-new
--replicas=3
deployment "nginx-deployment-new" scaled
```

读者可以自行验证，此处不再赘述。

注　　意

用户可以将 Pod 副本数缩容为 0，但是，在这种情况下 Deployment 本身不会被删除，只是可用副本数将会变成 0。

4.1.6 故障转移

Kubernetes 支持故障转移，保证集群中的各项服务的可用性。在 Kubernetes 中， Node 节点可动态增加到 Kubernetes 集群中，前提是这个节点已经正确安装、配置、启动了 kubelet 和 kube-proxy 等关键进程。默认情况下，kubelet 会向 Master 节点注册自己，这也是 Kubernetes 推荐的 Node 节点管理方式。一旦 Node 节点被纳入集群管理的范围，kubelet 会定时向 Master 节点汇报自身的情况，以及之前有哪些 Pod 在运行等，这样 Master 节点可以获知每个 Node 节点的资源使用情况，并实现高效均衡的资源调度策略。如果 Node 节点没有按时上报信息，则会被 Master 判断节点为失联，Node 节点状态会被标记为 Not Ready，随后 Master 节点会触发工作负载转移流程。

下面的内容模拟一次 Node 节点故障。首先确认 nginx-deployment-new 所创建的 Pod 有 4 个副本，其中 2 个副本运行在 127.0.0.1 节点上，2 个副本运行在 192.168.1.122 上，如下所示：

```
[root@localhost ~]# kubectl get pods -o wide
NAME                                 RE  ST       RES AGE IP              NODE
nginx-deployment-new-94093859-0qgkd     1/1  Running    0     7s  172.17.0.5
127.0.0.1
nginx-deployment-new-94093859-6d40m     1/1 Running     0     7s  172.17.0.4
192.168.1.122
nginx-deployment-new-94093859-kt7cf     1/1 Running     0     7s  172.17.0.5
192.168.1.122
nginx-deployment-new-94093859-xs3rs     1/1 Running     0     7s  172.17.0.6
127.0.0.1
```

然后将名为 192.168.1.122 的节点关闭，过一会儿，Master 节点会将 192.168.1.122 标记为 NotReady，如下所示：

```
[root@localhost ~]# kubectl get nodes
NAME            STATUS          AGE
127.0.0.1       Ready           1d
192.168.1.122       NotReady          16h
```

再等待一会，Master 节点会将原来运行在 192.168.1.122 上的 2 个副本标记为 Unknown 状态，然后在 127.0.0.1 节点上新创建 2 个副本，如下所示：

```
    [root@localhost ~]# kubectl get pods -o wide
    NAME                                RE  ST      RES AGE IP            NODE
    nginx-deployment-new-94093859-0qgkd   1/1 Running     0   14m 172.17.0.5
127.0.0.1
    nginx-deployment-new-94093859-6d40m   1/1 Unknown 0   14m 172.17.0.4
192.168.1.122
    nginx-deployment-new-94093859-9hxcn   1/1 Running 0   3m  172.17.0.9
127.0.0.1
    nginx-deployment-new-94093859-cldc0   1/1 Running     0   3m  172.17.0.10
127.0.0.1
    nginx-deployment-new-94093859-kt7cf   1/1 Unknown 0   14m 172.17.0.5
192.168.1.122
    nginx-deployment-new-94093859-xs3rs   1/1 Running 0   14m 172.17.0.6
127.0.0.1
```

即使 192.168.1.122 节点重新恢复正常，Master 节点也不会把 Pod 重新调度会该节点。

如果单个Pod副本出现故障，Kubernetes会自动创建一个新的副本，以达到预期的副本数。例如，我们使用 kubectl delete pod 命令将其中的一个副本删除，如下所示：

```
    [root@localhost ~]# kubectl delete pod nginx-deployment-new-94093859-xs3rs
    pod "nginx-deployment-new-94093859-xs3rs" deleted
```

稍等片刻，Kubernetes 会自动创建一个新的副本，并且调度到另外的节点上，如下所示：

```
    [root@localhost ~]# kubectl get pods -o wide
    NAME                                RE  ST      RES AGE IP           NODE
    nginx-deployment-new-94093859-0qgkd   1/1 Running     0   24m 172.17.0.5
      127.0.0.1
    nginx-deployment-new-94093859-9hxcn   1/1 Running     0   13m 172.17.0.9
127.0.0.1
    nginx-deployment-new-94093859-cldc0   1/1 Running     0   13m 172.17.0.10
127.0.0.1
    nginx-deployment-new-94093859-nsc84   1/1 Running     0   9s  172.17.0.2
192.168.1.122
```

4.1.7 通过标签控制 Pod 的位置

默认情况下，Kubernetes 会将 Pod 调度到所有可用的 Node 节点上，以达到负载平衡的目的。但是，在某些情况下，用户可以需要将一些特定功能的 Pod 部署在指定的 Node 节点上。例如，某些运行数据库的 Pod 需要运行在内存较大的 Node 节点上，而某些执行存储功能的 Pod 需要运行在磁盘 I/O 快的 Node 节点上。

在 Kubernetes 中，可以通过标签功能来标记各种资源，不只是节点，其他的资源，比如 Deployment、ReplicaSet 等都可以通过标签来标记。

所谓标签，实际上就是一些“键-值对”，在 Kubernetes 中，用户可以自由地定义自己的标签。

Kubernetes 通过 kubectl label node 命令来设置节点的标签。例如，下面的命令将名为 192.168.1.122 的节点设置一个名称为 mem 的标签：

```
[root@localhost ~]# kubectl label node 192.168.1.122 mem=large
node "192.168.1.122" labeled
```

通过上面的命令，标签已经被成功地添加到了节点上。用户可以通过--show-labels 选项将节点的标签显示出来，如下所示：

```
[root@localhost ~]# kubectl get node --show-labels
NAME            STATUS    AGE      LABELS
127.0.0.1       Ready     1d
beta.kubernetes.io/arch=amd64,beta.kubernetes.io/os=linux,kubernetes.io/hostn
ame=127.0.0.1
192.168.1.122   Ready     17h
beta.kubernetes.io/arch=amd64,beta.kubernetes.io/os=linux,kubernetes.io/hostn
ame=192.168.1.122,mem=large
```

从上面的输出结果可以看到，实际上 Kubernetes 已经自己维护了一些系统内部使用的标签，例如 beta.kubernetes.io/arch、beta.kubernetes.io/os 以及 kubernetes.io/hostname 等。

有了标签，我们就可以将 Pod 指定到特定的节点上了。Kubernetes 通过资源配置文件的 nodeSelector 属性来对节点进行选择。修改前面创建的 Deployment 配置文件，在 spec 属性中增加 nodeSelector 属性，如下所示：

```
apiVersion: extensions/v1beta1
kind: Deployment
metadata:
  name: nginx-deployment-new
spec:
  replicas: 3
  template:
    metadata:
      labels:
        app: nginx
        track: stable
    spec:
      containers:
      - name: nginx
        image: nginx:1.7.9
        ports:
```

```
            - containerPort: 80
      nodeSelector:
         mem: large
```

执行 kubectl apply 命令重新创建 Deployment，如下所示：

```
    [root@localhost ~]# kubectl apply -f nginx-deployment.yaml
    deployment "nginx-deployment-new" created
```

然后通过 kubectl get pods 命令查看 Pod 在节点上的调度情况：

```
    [root@localhost ~]# kubectl get pods -o wide
    NAME                              RE   ST        RES AGE IP          NODE
    nginx-deployment-new-3048195552-3c7     1/1 Running 0    19s 172.17.0.4
192.168.1.122
    nginx-deployment-new-3048195552-lt4     1/1 Running 0    19s 172.17.0.3
192.168.1.122
    nginx-deployment-new-3048195552-nnt     1/1 Running     0    19s 172.17.0.2
192.168.1.122
```

从上面的输出结果可知，现在所有的 Pod 都被调度到名为 192.168.1.122 节点上。

如果想要删除某个特定的标签，可以使用以下命令：

```
    [root@localhost ~]# kubectl label node 192.168.1.122 mem-
    node "192.168.1.122" labeled
```

在上面的命令中，mem-中的 mem 为标签的键，后面的减号表示将该标签删除。执行完成之后，再次查看节点的标签，就会发现标签已经被删除了，如下所示：

```
    [root@localhost ~]# kubectl get node --show-labels
    NAME            STATUS    AGE       LABELS
    127.0.0.1        Ready     1d        beta.kubernetes.io/arch=amd64,
beta.kubernetes.io/os=linux,kubernetes.io/hostname=127.0.0.1
    192.168.1.122   Ready      17h       beta.kubernetes.io/arch=amd64,
beta.kubernetes.io/os=linux,kubernetes.io/hostname=192.168.1.122
```

标签被删除后，Kubernetes 并不会立即重新调度 Pod 到其他节点上，除非用户通过 kubectl apply 命令重新部署：

```
    [root@localhost ~]# kubectl apply -f nginx-deployment.yaml
```

除了可以在配置文件中通过 nodeSelector 属性指定节点之外，用户在命令行中也可以通过-l 或者--labels=这两个选项指定节点的标签。例如，下面的命令也可以将 Pod 部署到含有标签 mem=large 的节点上：

```
    [root@localhost ~]# kubectl run nginx-deployment --image=nginx:1.7.9
--replicas=3 --labels='mem=large'
```

4.1.8 删除 Deployment

对于已经不再使用的 Deployment，用户将其从系统中删除。删除 Deployment 需要使用 kubectl delete deployment 命令。该命令的使用方法比较简单，直接将 Deployment 名作为参数传递给它就可以了，如下所示：

```
[root@localhost ~]# kubectl delete deployment nginx-deployment-new
deployment "nginx-deployment-new" deleted
```

当 Deployment 被删除之后，由该 Deployment 所创建的所有的 ReplicaSet 和 Pod 都会被自动删除。

4.1.9 DaemonSet

在前面的例子中，我们已经得知，通过 Deployment 创建的 Pod 副本会分布在各个节点上，每个节点上都有可能运行好几个副本。而 DaemonSet 则不同，通过它创建的 Pod 只会运行在同一个节点上，并且最多只会有一个副本。

DaemonSet 通常用来创建一些系统性的 Pod，例如提供日志管理、存储或者域名服务等。实际上，Kubernetes 本身就有一些 DaemonSet，这些 DaemonSet 都位于 kube-system 命名空间里面。例如，kube-proxy、kube-flannel-ds 等都是由 DaemonSet 创建的。

除了系统自身的 DaemonSet 之外，用户也可以自定义 DaemonSet。下面以一个简单的例子来说明如何运行自己的 DaemonSet。

Prometheus 的 Node Exporter 是一个非常有名的、开源的服务器监控系统。用户可以在 Kubernetes 集群中创建一个 Node Exporter 的 DaemonSet，用来监控各个容器的情况。

为了创建 DaemonSet，用户需要创建一个 YAML 配置文件，其名为 node-exporter-daemonset.yml，内容如下：

```
01 apiVersion: extensions/v1beta1
02 kind: DaemonSet
03 metadata:
04   name: node-exporter
05   namespace: monitoring
06   labels:
07     name: node-exporter
08 spec:
09   template:
10     metadata:
11       labels:
12         name: node-exporter
13       annotations:
14          prometheus.io/scrape: "true"
15          prometheus.io/port: "9100"
16     spec:
```

```
17        hostPID: true
18        hostIPC: true
19        hostNetwork: true
20        containers:
21          - ports:
22              - containerPort: 9100
23                protocol: TCP
24            resources:
25              requests:
26                cpu: 0.15
27            securityContext:
28              privileged: true
29            image: prom/node-exporter:v0.15.2
30            args:
31              - --path.procfs
32              - /host/proc
33              - --path.sysfs
34              - /host/sys
35              - --collector.filesystem.ignored-mount-points
36              - '"^/(sys|proc|dev|host|etc)($|/)"'
37            name: node-exporter
38            volumeMounts:
39              - name: dev
40                mountPath: /host/dev
41              - name: proc
42                mountPath: /host/proc
43              - name: sys
44                mountPath: /host/sys
45              - name: rootfs
46                mountPath: /rootfs
47        volumes:
48          - name: proc
49            hostPath:
50              path: /proc
51          - name: dev
52            hostPath:
53              path: /dev
54          - name: sys
55            hostPath:
56              path: /sys
57          - name: rootfs
58            hostPath:
59              path: /
```

在上面的配置文件中，第 2 行指定资源类型为 DaemonSet。第 5 行指定命名空间为 monitoring。第 19 行指定容器的网络模式为主机网络。第 29 行指定容器镜像为 prom/node-exporter:v0.15.2。第 32 行通过 Volume 将宿主机的/proc、/sys 以及/等路径映射到容器中。

由于命名空间 monitoring 在当前集群中并不存在，因此用户需要首先创建该命名空间，命令如下：

```
[root@localhost ~]# kubectl create namespace monitoring
```

创建完成之后，就可以创建 DaemonSet 了，如下所示：

```
[root@localhost ~]# kubectl create -f node-exporter-daemonset.yml
```

最后，用户可以通过 kubectl get pods 命令查看 Pod 状态，如下所示：

```
[root@localhost ~]# kubectl get pods --namespace=monitoring -o wide
NAME              RE   ST        RES AGE       IP              NODE
node-exporter-74c42     1/1 Running     0    7m       192.168.1.122
192.168.1.122
node-exporter-gjhfh     1/1 Running  0       7m       127.0.0.1
127.0.0.1
```

从上面的输出结果可知，在每个节点上都有一个副本在运行。

4.2 Job

在 4.1 节中，我们详细介绍了 Deployment 和 DaemonSet。在 Kubernetes 中，除了这 2 个重要的运行应用的途径之外，还有 Job。Job 在 Kubernetes 中的作用也是非常明显的，本节将详细介绍 Job 的使用方法。

4.2.1 什么是 Job

对于 Deployment、ReplicaSet 以及 Replication Controller 等类型的控制器而言，它希望 Pod 保持预期数目、持久地运行下去。除非用户明确删除，否则这些对象一直存在，它们针对的是持久性任务，如 Web 服务等。对于非持久性任务，比如归档文件，任务完成后，Pod 需要结束运行，不需要 Pod 继续保持在系统中，这个时候就要用到 Job。因此，可以说 Job 是对 ReplicaSet、Replication Controller 等持久性控制器的补充。

Job 的配置文件的语法与前面介绍的 Deployment 的基本语法大致相同。例如，下面的代码定义了一个 Job，该 Job 的功能是实现倒计时，其代码如下：

```
apiVersion: batch/v1
kind: Job
metadata:
  name: job-counter
spec:
  template:
```

```
    metadata:
      name: job-counter
    spec:
      restartPolicy: Never
      containers:
      - name: counter
        image: busybox
        command:
        - "bin/sh"
        - "-c"
        - "for i in 9 8 7 6 5 4 3 2 1; do echo $i; done"
```

在上面的代码中，kind 的值为 Job，表示当前创建的是一个 Job 类型的资源。restartPolicy 用来指定在什么情况下需要重启容器，在本例中的值为 Never。image 指定镜像为 busybox。Busybox 是一个集成了 100 多个最常用 Linux 命令和工具的软件工具箱，它在单一的可执行文件中提供了精简的 Unix 工具集。BusyBox 可运行于多款 POSIX 环境操作系统中，如 Linux（包括 Android）、Hurd、FreeBSD 等。Busybox 既包含了一些简单实用的工具，如 cat 和 echo，也包含了一些更大、更复杂的工具，如 grep、find、mount 以及 telnet，可以说 BusyBox 是 Linux 系统的瑞士军刀。

注　意

Job 的 restartPolicy 仅支持 Never 和 OnFailure 两种，不支持 Always，这是因为 Job 只相当于用来执行一个批处理任务，执行完就结束了。

将上面的代码保存为 job-counter.yaml，然后执行以下命令启动该 Job：

```
[root@localhost ~]# kubectl apply -f job-counter.yaml
```

然后通过 kubectl get jobs 命令查看 Job 的状态，如下所示：

```
[root@localhost ~]# kubectl get jobs
NAME          DESIRED    SUCCESSFUL    AGE
job-counter     1            1         25m
```

从上面命令的输出可知，job-counter 的预期值和成功值都是 1，这表示我们刚才已经成功执行了一个 Pod。

接下来，使用 kubectl get pods 命令查看 Pod 的状态，由于刚才启动的 Pod 已经退出运行，因此需要使用--show-all 选项，如下所示：

```
[root@localhost ~]# kubectl get pods --show-all
NAME                     READY     STATUS      RESTARTS    AGE
job-counter-6jkrf         0/1      Completed   0           26m
nginx-controller-8tw75    1/1      Running     0           5h
nginx-controller-t25nd    1/1      Running     0           5h
```

在上面的列表中，第1行的名为job-counter-6jkrf的Pod，正是刚刚执行完成的Job的Pod。可以得知，该Pod的状态为Completed，表示Pod已经执行完成。

上面的Pod的输出信息保存在日志中，用户可以通过kubectl logs命令查看，如下所示：

```
[root@localhost ~]# kubectl logs job-counter-6jkrf
9
8
7
6
5
4
3
2
1
[root@localhost ~]#
```

这个Job已经按照我们的预期，输出了从9~1的倒计时序列。

4.2.2 Job失败处理

作为一种任务计划，总有在执行过程中遇到某些意外的时候。为了能够及时处理这种意外，用户需要及时掌握Job的执行情况。

为了演示Job执行失败的情况，我们修改job-counter.yaml配置文件，将其中的command的值修改为一个错误的命令，如下所示：

```
apiVersion: batch/v1
kind: Job
metadata:
  name: job-counter
spec:
  template:
    metadata:
      name: job-counter
    spec:
      restartPolicy: Never
      containers:
      - name: counter
        image: busybox
        command: ["bad command"]
```

然后删除前面创建的Job：

```
[root@localhost ~]# kubectl delete job job-counter
job "job-counter" deleted
```

接下来创建新的 Job：

```
[root@localhost ~]# kubectl apply -f job-counter.yaml
job "job-counter" created
```

执行完成之后，通过 kubectl get jobs 命令查看 Job 信息：

```
[root@localhost ~]# kubectl get jobs
NAME           DESIRED       SUCCESSFUL       AGE
job-counter        1              0           1m
```

从上面的输出结果可以得知，预期的副本数为 1，而成功的副本数为 0。

如果使用 kubectl get pods 命令查看 Pod 状态，其结果如下：

```
[root@localhost ~]# kubectl get pods --show-all
NAME                 READY        STATUS              RESTARTS   AGE
job-counter-0ds5m      0/1          ContainerCannotRun       0           2m
job-counter-3b4x5      0/1          ImagePullBackOff         0           1m
job-counter-6gpgd      0/1          ContainerCannotRun       0           2m
job-counter-7xmkx     0/1           ContainerCannotRun       0           1m
job-counter-dks22      0/1          ContainerCannotRun       0           2m
job-counter-fq7cf      0/1          ContainerCannotRun       0           3m
job-counter-fz83p       0/1         ContainerCannotRun       0           2m
job-counter-g7749      0/1          ContainerCannotRun       0           1m
job-counter-h0gmp      0/1          ContainerCannotRun       0           3m
job-counter-hqw80     0/1           ContainerCannotRun       0           1m
job-counter-nggwv      0/1          ContainerCannotRun       0           3m
job-counter-q360q      0/1          ContainerCannotRun       0           3m
job-counter-rs2bs       0/1         ContainerCannotRun       0           2m
job-counter-s5qsj       0/1         ContainerCannotRun       0           1m
job-counter-skk9t       0/1         ContainerCannotRun       0           1m
job-counter-tcdp0       0/1         ContainerCannotRun       0           1m
job-counter-w5bgc      0/1          ContainerCannotRun       0           2m
```

可以发现，上面的输出结果中存在着多个状态为 ContainerCannotRun 的 Pod。从 Pod 名称中的前缀可以得知，这些 Pod 都是由前面的 Job 创建的。此时，如果使用 kubectl describe pod 命令随便查看一个 Pod 的信息，则会发现以下错误信息：

```
    14m          14m               1        {kubelet 127.0.0.1}
spec.containers{counter}          Warning          Failed                 Failed to
start container with docker id 457639ebae4e with error: Error response from daemon:
{"message":"oci runtime error: container_linux.go:247: starting container process
caused \"exec: \\\"bad command\\\": executable file not found in $PATH\"\n"}
```

上面的错误日志表明，由于我们在配置文件中指定的命令没有被找到，从而导致 Job 运行失败。

但是读者可能会问，为什么会出现那么多的 Pod 呢？其原因在于，尽管我们在配置文件中将 restartPolicy 的值指定为 Never，这样在 Job 启动失败之后，容器不会被重新启动，但是由于预期 Pod 的副本数为 1，而目前的副本数却一直是 0，达不到预期的数量。此时，Job 会不停地创建新的 Pod，一直到达到成功的副本数为 1 为止。

为了终止这种情况，我们需要删除该 Job，命令如下：

```
[root@localhost ~]# kubectl delete -f job-counter.yaml
```

4.2.3 Job 的并行执行

在某些情况下，用户可能需要同时运行多个 Job 副本，以提高工作效率。此时，就涉及某个 Job 的多个副本并行执行的问题。

Job 的并行执行需要使用 parallelism 属性，该属性为数值型，默认情况下，该属性的值为 1，表示只有一个副本在执行。如果需要多个副本同时执行，就需要修改该属性的值。

我们将 4.2.1 小节中的 Job 的配置文件进行修改，增加 parallelism 属性，将其值设置为 3，如下所示：

```
apiVersion: batch/v1
kind: Job
metadata:
  name: job-counter
spec:
  parallelism: 3
  template:
    metadata:
      name: job-counter
    spec:
      restartPolicy: Never
      containers:
      - name: counter
        image: busybox
        command:
        - "bin/sh"
        - "-c"
        - "for i in 9 8 7 6 5 4 3 2 1; do echo $i; done"
```

然后通过以下命令运行 Job：

```
[root@localhost ~]# kubectl apply -f job-counter.yaml
```

执行完成之后，查看 Job 状态，如下所示：

```
[root@localhost ~]# kubectl get jobs
NAME            DESIRED          SUCCESSFUL          AGE
job-counter         <none>           3                   2m
```

通过上面的输出可以得知，一共有 3 个 Job 成功执行。接下来查看以下 Pod 的状态，如下所示：

```
[root@localhost ~]# kubectl get pods --show-all
NAME                 READY     STATUS      RESTARTS     AGE
job-counter-2w53g     0/1      Completed   0            2m
job-counter-4vvxl     0/1      Completed   0            2m
job-counter-r1bjs     0/1      Completed   0            2m
```

从上面的输出结果可知，一共有 3 个 Pod 处于完成状态，它们的 AGE 值都为 2m，这表示它们是同时并行执行的。

4.2.4 Job 定时执行

在 Kubernetes 中，Job 除了可以并行执行之外，还可以定时执行。这在完成某些计划任务时非常重要。Kubernetes 为定时 Job 专门设置了一种资源类型，其名为 CronJob。

CrontabJob 需要 batch/v2alpha1 版本的 API，所以需要预先修改 kube-apiserver 的配置文件，增加以下选项：

```
KUBE_API_ARGS="--allow_privileged=true
--runtime-config=batch/v2alpha1=true"
```

然后创建一个 CronJob 的 YAML 配置文件，代码如下：

```
apiVersion: batch/v1
kind: CronJob
metadata:
  name: hello
spec:
  schedule: "*/1 * * * *"
  jobTemplate:
    spec:
      template:
        spec:
          containers:
          - name: hello
            image: busybox
            command: ["echo","hello k8s job!"]
          restartPolicy: OnFailure
```

执行 kubectl apply 命令，创建 CronJob，如下所示：

```
[root@localhost ~]# kubectl apply -f cronjob.yaml
```

然后查看 CronJob 的状态：

```
[root@localhost ~]# kubectl get cronjob hello
NAME        SCHEDULE        SUSPEND    ACTIVE      LAST-SCHEDULE
hello       */1 * * * *        False        0            <none>
```

可以发现，在上面的列表中，没有 Job 处于活动状态。为了能够观察到 Job 的执行情况，需要使用--watch 选项，如下所示：

```
[root@localhost ~]# kubectl get jobs --watch
NAME               DESIRED     SUCCESSFUL       AGE
hello-4111706356     1            1             2s
```

现在，从上面的输出结果可以发现，每隔 1 分钟左右，上面的 Job 会执行一次。

除了使用配置文件之外，用户也可以使用命令行来创建 CronJob。例如，上面的 CronJob 可以使用以下命令来创建：

```
[root@localhost ~]# kubectl run hello --schedule="*/1 * * * *"
--restart=OnFailure --image=busybox -- /bin/sh -c "date; echo Hello from the
Kubernetes cluster"
cronjob "hello" created
```

第 5 章

通过服务访问应用

在第 4 章中，我们详细介绍了在 Kubernetes 中如何运行一个应用。Pod 是 Kubernetes 运行应用的基础，也就是说 Kubernetes 的各项服务都是由 Pod 提供的。作为一个用户来说，当然希望自己的应用是健壮的。但是，Pod 本身并不是健壮的，这就给用户带来一个困惑，那就是如何使自己的各种服务变得健壮起来。本章将对这个问题进行详细讨论。

本章涉及的知识点有：

- 服务及其功能：主要介绍 Kubernetes 中服务的基本概念及其用途。
- 管理服务：介绍服务的创建、查看以及删除等常见的操作。
- 外网访问服务：主要介绍 ClusterIP、NodePort 和 LoadBalancer 这三种方式的使用场景。
- 通过 Cluster DNS 访问服务：主要介绍 Cluster DNS 的安装、配置以及如何通过 DNS 访问服务。

5.1 服务及其功能

服务是 Kubernetes 系统体系中的一个核心概念。借助服务，用户可以非常方便地实现应用的服务发现与负载均衡，并实现应用的零宕机升级。本节将详细介绍服务的概念及其功能，使读者能够深入掌握这个概念。

5.1.1 服务基本概念

通过前面一章的学习，读者可能会了解到，Kubernetes 中的 Pod 是有生命周期的，它们可以被创建，也可以被销毁，然而一旦被销毁生命就永远结束。用户通过 ReplicaSets 能够动态地创建和销毁 Pod，比如需要进行扩缩容，或者执行滚动升级。另外，Pod 本身也会产生故障，发生意外退出的情况。在这种情况下，Kubernetes 会自动创建一个新的 Pod 副本来代替故障 Pod。

每当新的 Pod 被创建时，它都会获取它自己的 IP 地址，这意味着 Pod 的 IP 地址并不总是稳定可依赖的。这会导致一个问题，在 Kubernetes 集群中，如果一组运行后台服务的 Pod 为其他运行前台服务的 Pod 提供服务，那么那些提供前台服务的 Pod 该如何发现，并连接到这组提供后台服务的 Pod 上呢？

Kubernetes 的 Service 就是解决这个问题而提出的。它定义了这样一种抽象为逻辑上的一组 Pod，提供了一种可以访问它们的策略，这种策略通常被称为微服务。这组 Pod 能够被 Service 访问到。

举个例子，一个图片处理后台服务应用，它运行了 3 个 Pod 副本。这些副本是可互换的，前台应用不需要关心它们具体调用了哪个后台 Pod 副本。然而组成这一组后台图片处理程序的 Pod 实际上可能会发生变化，前台客户端不应该也没必要知道，而且也不需要跟踪这一组后台服务的状态，这些工作由 Service 来完成，前台服务只要关心相应的服务是否正常服务就可以了。服务定义的抽象能够解耦这种 Pod 之间的关联。

5.1.2　服务的功能原理

服务的主要功能有两个，一个是实现了服务之间的关联解耦，另外一个就是实现了服务发现和负载均衡。图 5-1 所示描述了一个服务与一组 Pod 的关系。

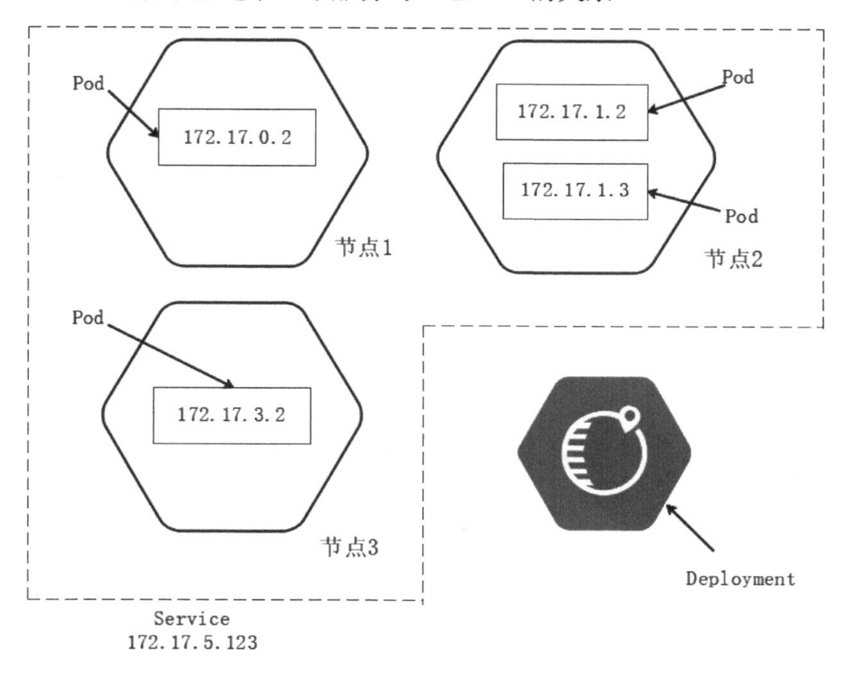

图 5-1　一个服务与一组 Pod 的关系

图 5-1 中黑色六边形是一个节点，节点可以是一台主机或者虚拟机。虚线是由三个节点组成的服务，提供负载均衡和服务发现，有一个固定的 IP 地址 172.17.5.123。长方形为 Pod，Pod 的 IP 地址是不固定的，因为需要经常生成和销毁。

在逻辑层面上，服务被认为是真实应用的抽象，每一个服务通过标签选择器关联着一系列的 Pod。在物理层面上，服务又是用户应用的代理服务器，对外表现为一个单一访问入口，通过 Kubernetes Proxy 转发请求到服务关联的 Pod。

在应用的滚动更新过程中，Kubernetes 通过逐个容器替代升级的方式实现了无中断的服务升级，如图 5-2 所示。

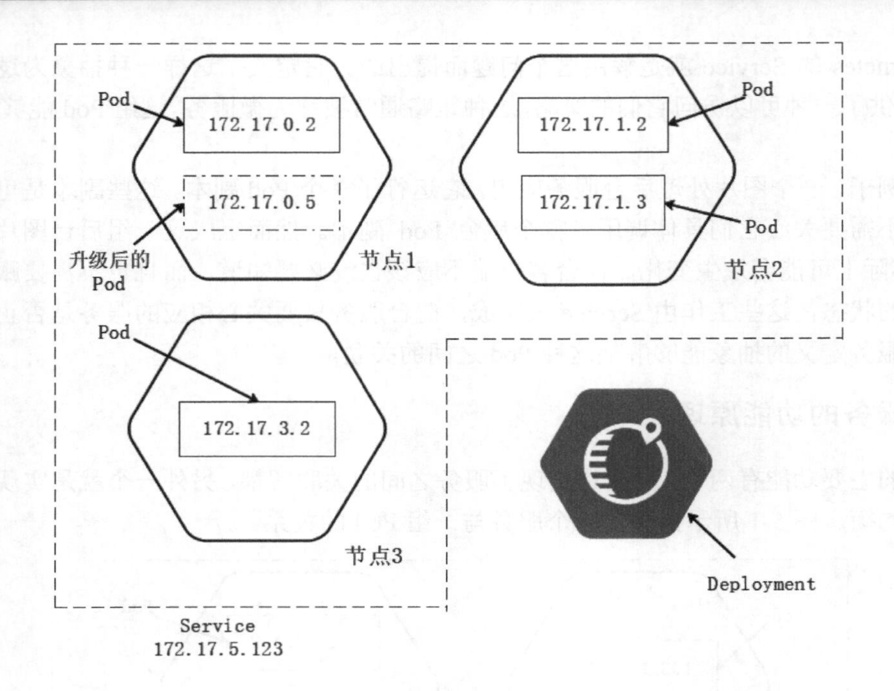

图 5-2　滚动更新

在图 5-2 中，首先更新的是左上角的节点。节点中实线长方形为原来的 Pod，虚线长方形为更新后的 Pod。更新的过程中会创建一个新的 Pod，并且拥有新的 IP 地址 172.17.0.5。当新的 Pod 处于可用状态后，旧的 Pod 便被销毁，而其他节点的 Pod 也会依次开始更新。而从服务的服务对象角度来看，整个过程中服务并没有发生变化，所以服务可以一直保持服务。

5.2 管 理 服 务

跟其他的资源一样，服务的管理主要包括创建、查看以及删除等操作。本节将对这些常见的操作进行介绍。

5.2.1　创建服务

同样，服务也是 Kubernetes 中的一种资源，所以它的配置文件与前面介绍的 YAML 配置文件非常相似。

一个典型的服务的 YAML 配置文件如下：

```
01  apiVersion: v1
02  kind: Service
03  matadata:
04    name: string
05    namespace: string
06    labels:
07      - name: string
```

```
08   annotations:
09     - name: string
10  spec:
11   selector: []
12   type: string
13   clusterIP: string
14   sessionAffinity: string
15   ports:
16   - name: string
17     protocol: string
18     port: int
19     targetPort: int
20     nodePort: int
21   status:
22     loadBalancer:
23       ingress:
24         ip: string
25         hostname: string
```

第 1 行的 apiVersion 为 API 的版本号，这个是必需属性。第 2 行的 kind 表示该项资源为服务。第 3~9 行定义 Service 的元数据，其中 name 表示服务的名称，为必需属性。从第 10 行开始，定义 Service 的规格，其中绝大部分属性都是非必需的。需要注意的是，第 11 行的 selector 为标签选择器，用来选择服务可以代理的 Pod。第 12 行的 type 用来标识服务的类型，可以是以下三种类型：

- ClusterIP：该项为默认的类型，提供一个集群内部的虚拟 IP，与 Pod 不在同一网段，以供集群内部的 Pod 之间通信使用。
- nodePort：使用宿主机的端口，是能够访问各节点的外部客户通过节点的 IP 地址和端口就能访问服务。
- loadbalancer：使用外接负载均衡完成服务到负载的分发，需要在 spec.status.loadBalancer 字段指定负载均衡器的 IP 地址，并同时定义 nodePort 和 ClusterIP。

下面通过配置文件创建一个 Service，配置文件的内容如下：

```
apiVersion: v1
kind: Service
metadata:
  name: web-service
spec:
  selector:
    app: webserver
  ports:
  - name: http
    protocol: TCP
```

```
    port: 80
    targetPort: 80
  - name: https
    protocol: TCP
    port: 443
    targetPort: 443
```

将以上配置文件保存为 service .yaml，然后执行以下命令创建服务：

```
kubectl create -f service.yaml
```

除了使用配置文件之外，用户还可以通过命令行创建 Service。命令的基本语法如下：

```
kubectl create service
```

通过命令，用户可以创建 4 种类型的服务，分别为 ClusterIP、LoadBalancer、nodeport 以及 ExternalName。其中，ClusterIP 的创建命令的基本语法如下：

```
kubectl create clusterip name [--tcp=<port>:<targetport>]
```

这个命令中，name 为服务的名称。--tcp 选项用来指定 Service 的服务端口与 Pod 端口之间的映射，其中 port 为 Service 对外提供服务的端口，targetport 为服务关联的 Pod 的端口。对于这两个端口，读者一定需要搞清楚它们的用途。简单地说，port 就是 Service 对应的 ClusterIP 的 IP 地址。当其他的应用访问 port 所代表的端口时，Service 就将该请求转发到 targetport 所定义的端口，进而被转发到 Pod 的相应端口。在本例中，当其他应用访问 80 端口时，就被转发到 Pod 的 80 端口，即 Pod 中 Nginx 的服务端口。同样，当其他应用访问 HTTPS 的 443 端口时，请求就会被转发到 Pod 的 443 端口，即 Nginx 提供 HTTPS 服务的端口。

例如，下面的命令创建一个名为 myservice 的 clusterip，并且将服务端口映射到 8080 端口：

```
[root@localhost ~]# kubectl create service clusterip my-cs --tcp=5678:8080
```

创建 LoadBalancer 的命令的基本语法如下：

```
kubectl create loadbalancer name [--tcp=port:targetport]
```

创建 nodeport 的命令的基本语法如下：

```
kubectl create nodeport name [--tcp=port:targetport]
```

创建 ExternalName 的命令的基本语法如下：

```
kubectl create externalname name --external-name external.name
```

这些命令的使用方法与 clusterIP 基本相同，不再举例说明。

5.2.2 查看服务

用户可以通过命令 kubectl get svc 或者 kubectl get services 查看创建的服务，如下所示：

```
[root@localhost ~]# kubectl get svc
NAME          CLUSTER-IP          EXTERNAL-IP     PORT(S)          AGE
```

```
kubernetes     10.254.0.1      <none>       443/TCP           4d
my-cs          10.254.75.70    <none>       5678/TCP          16m
web-service    10.254.250.54   <none>       80/TCP,443/TCP    48m
```

在上面的输出结果中，NAME 为 Service 的名称，CLUSTER-IP 为 Kubernetes 给服务分配的 IP 地址。EXTERNAL-IP 为外部 IP 地址，如果没有指定，则为 none。PORTS 为服务的服务端口。

如果想要显示更加详细的信息，可以使用-o wide 选项，如下所示：

```
[root@localhost ~]# kubectl get svc -o wide
NAME         CLUSTER-IP      EXTERNAL-IP  PORT(S)          AGE    SELECTOR
kubernetes   10.254.0.1      <none>       443/TCP          4d     <none>
my-cs        10.254.75.70    <none>       5678/TCP         17m    app=my-cs
web-service  10.254.250.54   <none>       80/TCP,443/TCP   49m    name=nginx
```

在上面的输出结果中，SELECTOR 即为创建 Service 时指定的标签选择器。

除了查看服务状态信息之外，用户还可以通过 kubectl get endpoints 命令查看服务代理的 Pod 的信息，如下所示：

```
[root@localhost ~]# kubectl get endpoints web-service
NAME          ENDPOINTS                                                    AGE
web-service   172.17.0.2:443,172.17.0.2:443,172.17.0.3:80 + 1 more...     1h
```

其中 ENDPOINTS 为当前服务所关联的 Pod 的 IP 地址及其端口。

如果想要查看更为详细的关于服务的信息，则可以使用 kube describe service 命令：

```
[root@localhost ~]# kubectl describe service web-service
Name:                      web-service
Namespace:                 default
Labels:                    <none>
Selector:                  name=nginx
Type:                      ClusterIP
IP:                        10.254.123.250
Port:                      http    80/TCP
Endpoints:                 172.17.0.2:80,172.17.0.3:80
Port:                      https   443/TCP
Endpoints:                 172.17.0.2:443,172.17.0.3:443
Session Affinity:          None
No events.
```

从上面的输出可知，所使用的选择器为 name=nginx，Service 的类型为 ClusterIP，端口 80 所关联的 Endpoints 分别为 172.17.0.2:80 和 172.17.0.3:80，端口 433 管理的 Endpoints 分别为 172.17.0.2:443 和 172.17.0.3:443。

5.2.3 销毁服务

销毁服务有两种方式。如果用户是通过 YAML 配置文件创建的服务，则可以使用 kubectl delete -f 命令将服务删除。例如，我们想要删除刚才创建的 web-service，命令如下：

```
[root@localhost ~]# kubectl delete -f service.yaml
service "web-service" deleted
```

而对于命令行创建的服务，由于没有对应的配置文件，因此无法使用上面的命令将其删除。与之相对应，Kubernetes 也提供了命令行的删除方法。例如，我们想要将名为 my-cs 的服务删除，则命令如下：

```
[root@localhost ~]# kubectl delete service my-cs
service "my-cs" deleted
```

5.3 外部网络访问服务

服务想要对外提供服务，就必须能够被外部网络的各种信息系统所访问。在 Kubernetes 中，用户可以通过很多种方式来达到这个目的。本节将对其中最常用的两种方式进行介绍。

5.3.1 kube-proxy 结合 ClusterIP

前面已经讲过，服务就是一组 Pod 的服务抽象，相当于一组 Pod 的负载均衡，负责将请求分发给对应的 Pod。服务会为这个负载均衡提供一个 IP 地址，一般称为 ClusterIP。ClusterIP 是一个虚拟的 IP 地址，是由 Kubernetes 分配给服务使用的。

kube-proxy 的作用主要是负责服务的实现。具体来说，就是实现了内部从 Pod 到服务和外部的从 NodePort 向服务的访问，如图 5-3、图 5-4 所示。

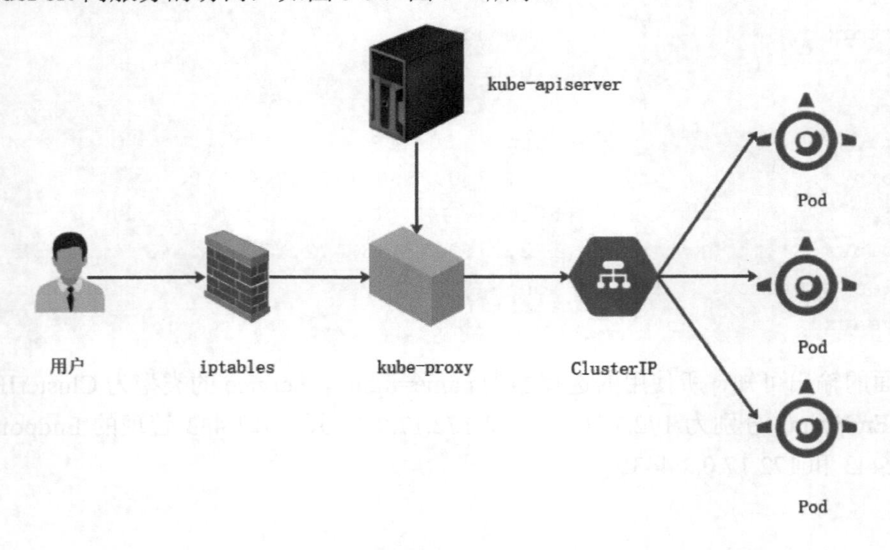

图 5-3 kube-proxy

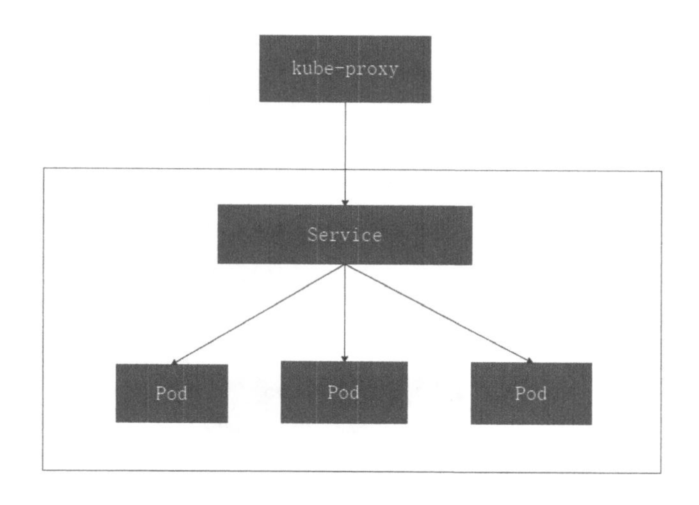

图 5-4 通过 kube-proxy 访问 ClusterIP 类型的服务（Service）

Kubernetes 的 API 提供了外部访问 ClusterIP 的能力，其访问方式如下：

```
    http://masterip:8080/api/v1/proxy/namespaces/<namespace>/services/<servic
e-name>:<port-name>/
```

在上面的语法中，masterip 为 Master 节点的 IP 地址，<namespace>为 Service 所属的命名空间，<service-name>为 Service 的名称，<port-name>为端口。

例如，对于前面定义的名为 web-service 的 Service，我们可以通过以下 URL 来访问：

http://192.168.1.121:8080/api/v1/proxy/namespaces/default/services/web-service:http/

访问结果如图 5-5 所示。

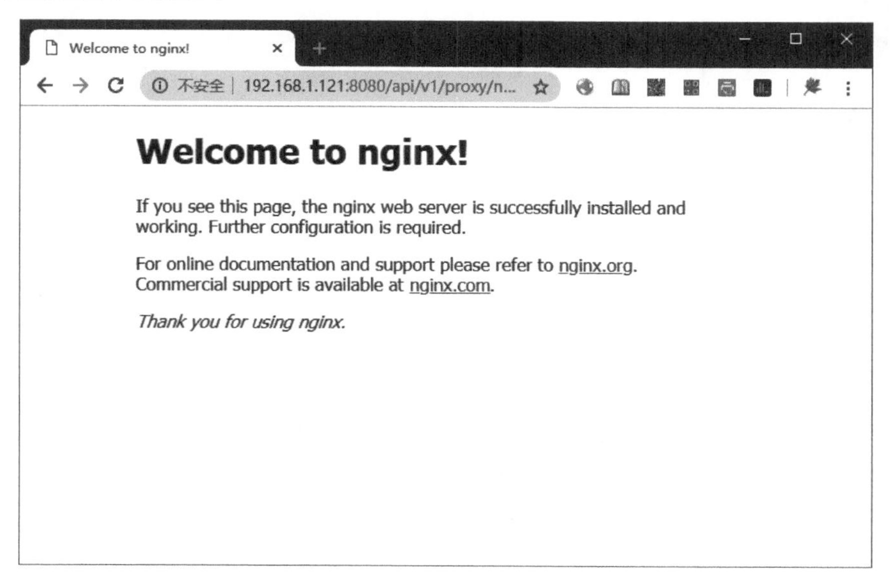

图 5-5 通过 kube-proxy 访问服务（Service）

5.3.2 通过 NodePort

NodePort类型的服务是让外部机器可以访问集群内部服务的最基本的方式。所谓NodePort，顾名思义可以在所有节点上打开一个特定端口，任何发送到此端口的流量都将转发到相应的服务上面。图 5-6 所示描述了通过 NodePort 访问服务的原理。

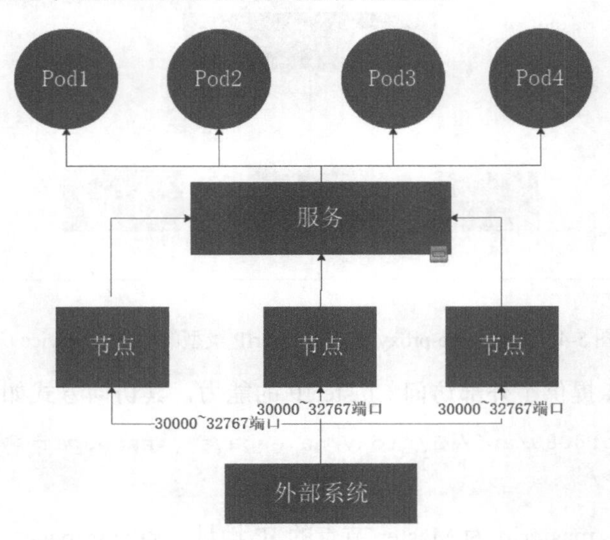

图 5-6　通过 NodePort 访问服务

例如，下面的代码为一个 NodePort 类型的服务的 YAML 配置文件：

```
apiVersion: v1
kind: Service
metadata:
  name: nginx-service-nodeport
spec:
  selector:
      app: nginx
  ports:
    - name: http
      port: 8000
      protocol: TCP
      targetPort: 80
  type: NodePort
```

将以上代码保存为 nodeport.yaml，然后使用以下命令创建服务：

```
[root@localhost ~]# kubectl apply -f nodeport.yaml
service "nginx-service-nodeport" created
```

查看服务状态，如下所示：

```
[root@localhost ~]# kubectl get svc -o wide
NAME                  CLUSTER-IP      EXTERNAL-IP   PORT(S)      AGE   SELECTOR
kubernetes            10.254.0.1      <none>        443/TCP      32d   <none>
my-nginx              10.254.27.146   <none>        80/TCP       14m   app=nginx
nginx-service-nodeport 10.254.112.42   <nodes>   8000:30243/TCP 9s
app=nginx
```

在上面的输出中，最后一行即为刚刚创建的 NodePort 类型的服务。可以得知，该服务将 8000 端口映射到了节点的 30243 端口。而在上面的定义中，服务的目标端口为 Pod 的 80 端口。因此，用户可以通过访问节点的 30243 端口，间接地访问 Pod 的 80 端口，如图 5-7 所示。

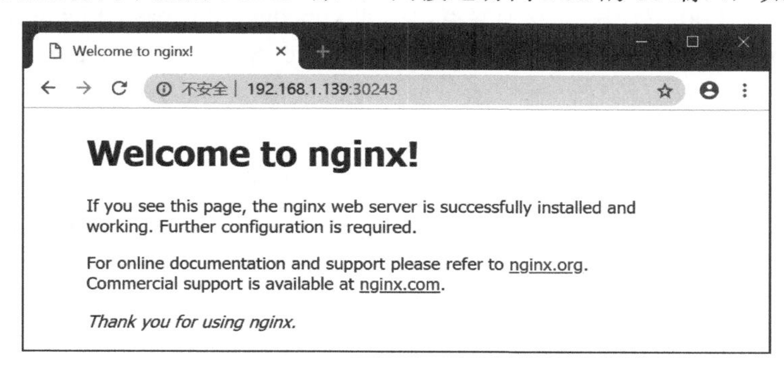

图 5-7 通过 NodePort 访问服务

5.3.3 通过负载均衡

负载均衡类型的服务是在公网上面发布服务的标准方式。例如，用户在云或者本地的机器上面有个负载均衡系统。该负载均衡系统通常会有一个固定的 IP 地址。当外部网络的其他系统访问该 IP 地址时，会将请求转发到服务上面，其原理如图 5-8 所示。

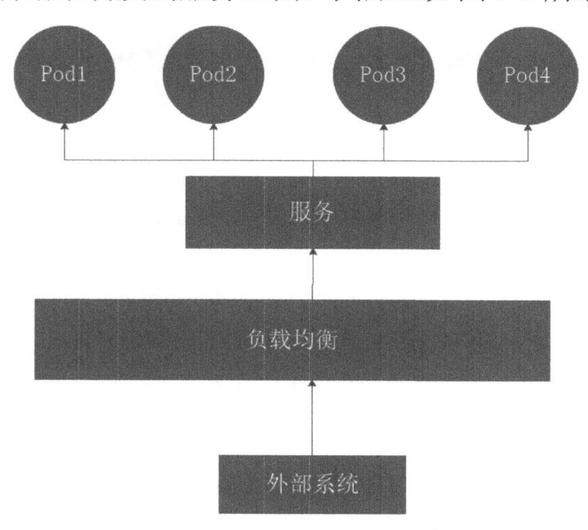

图 5-8 通过负载均衡访问服务

5.4 通过 CoreDNS 访问应用

前面已经介绍了多种通过服务来访问应用的方法。这些方法在一定程度上解决了 Pod 重建后 IP 动态变化以及负载均衡问题，但使用服务还是要需要预先知道服务的 ClusterIP，因此存在着一定的局限性。CoreDNS 就是专门为了解决这个问题而提出的，本节将详细介绍通过 CoreDNS 来实现服务发现的方法。

5.4.1 CoreDNS 简介

在早期版本的 Kubernetes 中，使用 kube-dns 插件来实现基于域名服务（DNS）的服务发现。kube-dns 在一个 Pod 中使用了多个容器，例如 kubedns、dnsmasq 和 sidecar。kubedns 容器监视 Kubernetes API，并基于 Kubernetes DNS 规范提供 DNS 记录；dnsmasq 提供缓存和存根域支持；sidecar 提供指标和健康检查。由于涉及多个容器，因此在实际运行过程中，kube-dns 会出现一些难以避免的问题。

CoreDNS 是一个通用的、权威的 DNS 服务器，提供与 Kubernetes 后向兼容但可扩展的集成。它解决了 kube-dns 所遇到的问题，并提供了许多独特的功能，可以解决各种各样的问题。与 kube-dns 不同，在 CoreDNS 中，所有这些功能都在一个容器中完成。

5.4.2 安装 CoreDNS

在 Kubernetes 中部署 CoreDNS 作为集群内的 DNS 服务有很多种方式，例如可以使用官方的软件包管理工具 Helm 部署，也可以通过 YAML 配置文件进行部署。为了能够使读者深入了解 CoreDNS 的部署过程，我们讲解一下通过 YAML 文件的部署方法。

首先是获取官方的 CoreDNS 的 YAML 配置文件代码，用户可以直接在 GitHub 的网站上面下载，其网址为：https://github.com/kubernetes/kubernetes/tree/master/cluster/addons/dns/coredns。也可以通过下载二进制包 kubernetes-server-linux-amd64.tar.gz，从该压缩包中得到，命令如下：

```
[root@localhost ~]# tar -zxvf kubernetes-server-linux-amd64.tar.gz
```

然后进入 kubernetes 目录，会发现一个名为 kubernetes-src.tar.gz 的 Kubernetes 源代码文件。通过以下命令解压该文件：

```
[root@localhost kubernetes]# tar -zxvf kubernetes-src.tar.gz
```

其中 CoreDNS 的 YAML 配置文件位于 cluster/addons/dns/coredns 目录中，如下所示：

```
[root@localhost coredns]# ll
total 36
-rw-rw-r--    1   root    root    4240    Feb 27 15:10
coredns.yaml.base
-rw-rw-r--    1   root    root    4308    Feb 27 15:10    coredns.yaml.in
-rw-rw-r--    1   root    root    4281    Feb 27 15:10
coredns.yaml.sed
```

```
-rw-rw-r--      1   root    root    1075    Feb 27 15:10    Makefile
-rw-rw-r--      1   root    root    308     Feb 27 15:10
transforms2salt.sed
-rw-rw-r--      1   root    root    266     Feb 27 15:10
transforms2sed.sed
```

在上面的文件列表中，coredns.yaml.base 就是我们所需要的 YAML 模板文件。

将上述 YAML 模板文件另存为 coredns.yaml，命令如下：

```
[root@localhost coredns]# cp coredns.yaml.base coredns.yaml
```

然后对 coredns.yaml 进行相应的修改，修改后的完整代码如下：

```
01 apiVersion: v1
02 kind: ServiceAccount
03 metadata:
04   name: coredns
05   namespace: kube-system
06 ---
07 apiVersion: rbac.authorization.k8s.io/v1alpha1
08 kind: ClusterRole
09 metadata:
10   labels:
11     kubernetes.io/bootstrapping: rbac-defaults
12   name: system:coredns
13 rules:
14 - apiGroups:
15   - ""
16   resources:
17   - endpoints
18   - services
19   - pods
20   - namespaces
21   verbs:
22   - list
23   - watch
24 - apiGroups:
25   - ""
26   resources:
27   - nodes
28   verbs:
29   - get
30 ---
31 apiVersion: rbac.authorization.k8s.io/ v1alpha1
32 kind: ClusterRoleBinding
```

```
33 metadata:
34   annotations:
35     rbac.authorization.kubernetes.io/autoupdate: "true"
36   labels:
37     kubernetes.io/bootstrapping: rbac-defaults
38   name: system:coredns
39 roleRef:
40   apiGroup: rbac.authorization.k8s.io
41   kind: ClusterRole
42   name: system:coredns
43 subjects:
44 - kind: ServiceAccount
45   name: coredns
46   namespace: kube-system
47 ---
48 apiVersion: v1
49 kind: ConfigMap
50 metadata:
51   name: coredns
52   namespace: kube-system
53 data:
54   Corefile: |
55     .:53 {
56         errors
57         health
58         ready
59         kubernetes cluster.local in-addr.arpa ip6.arpa {
60           pods insecure
61           fallthrough in-addr.arpa ip6.arpa
62         }
63         prometheus :9153
64         forward . /etc/resolv.conf
65         cache 30
66         loop
67         reload
68         loadbalance
69     }
70 ---
71 apiVersion: extensions/v1beta1
72 kind: Deployment
73 metadata:
74   name: coredns
75   namespace: kube-system
```

```
76    labels:
77      k8s-app: kube-dns
78      kubernetes.io/name: "CoreDNS"
79  spec:
80    replicas: 2
81    strategy:
82      type: RollingUpdate
83      rollingUpdate:
84        maxUnavailable: 1
85    selector:
86      matchLabels:
87        k8s-app: kube-dns
88    template:
89      metadata:
90        labels:
91          k8s-app: kube-dns
92      spec:
93        priorityClassName: system-cluster-critical
94        serviceAccountName: coredns
95        tolerations:
96          - key: "CriticalAddonsOnly"
97            operator: "Exists"
98        nodeSelector:
99          beta.kubernetes.io/os: linux
100       containers:
101       - name: coredns
102         image: coredns/coredns:1.5.0
103         imagePullPolicy: IfNotPresent
104         resources:
105           limits:
106             memory: 170Mi
107           requests:
108             cpu: 100m
109             memory: 70Mi
110         args: [ "-conf", "/etc/coredns/Corefile" ]
111         volumeMounts:
112         - name: config-volume
113           mountPath: /etc/coredns
114           readOnly: true
115         ports:
116         - containerPort: 53
117           name: dns
118           protocol: UDP
```

```
119             - containerPort: 53
120               name: dns-tcp
121               protocol: TCP
122             - containerPort: 9153
123               name: metrics
124               protocol: TCP
125             securityContext:
126               allowPrivilegeEscalation: false
127               capabilities:
128                 add:
129                 - NET_BIND_SERVICE
130                 drop:
131                 - all
132               readOnlyRootFilesystem: true
133             livenessProbe:
134               httpGet:
135                 path: /health
136                 port: 8080
137                 scheme: HTTP
138               initialDelaySeconds: 60
139               timeoutSeconds: 5
140               successThreshold: 1
141               failureThreshold: 5
142             readinessProbe:
143               httpGet:
144                 path: /ready
145                 port: 8181
146                 scheme: HTTP
147           dnsPolicy: Default
148           volumes:
149             - name: config-volume
150               configMap:
151                 name: coredns
152                 items:
153                 - key: Corefile
154                   path: Corefile
155     ---
156     apiVersion: v1
157     kind: Service
158     metadata:
159       name: kube-dns
160       namespace: kube-system
161       annotations:
```

```
162      prometheus.io/port: "9153"
163      prometheus.io/scrape: "true"
164    labels:
165      k8s-app: kube-dns
166      kubernetes.io/cluster-service: "true"
167      kubernetes.io/name: "CoreDNS"
168  spec:
169    selector:
170      k8s-app: kube-dns
171    clusterIP: 10.254.0.2
172    ports:
173    - name: dns
174      port: 53
175      protocol: UDP
176    - name: dns-tcp
177      port: 53
178      protocol: TCP
179    - name: metrics
180      port: 9153
181      protocol: TCP
```

其中，第 2~5 行创建 ServiceAccount；第 7~29 行创建 ClusterRole，其名称为 system:coredns；第 31~46 行创建 ClusterRoleBinding 对象；第 48~69 行创建 ConfigMap 对象；第 71~154 行创建 Deployment 对象；第 157~181 行创建服务对象，其中第 171 行指定所创建的服务的 ClusterIP 为 10.254.0.2。

使用以下命令部署 CoreDNS：

```
[root@localhost coredns]# kubectl create -f coredns.yml
```

部署完成之后，查看 Pod 是否部署成功，如下所示：

```
[root@localhost coredns]# kubectl get pod --namespace=kube-system
NAME                                    READY   STATUS    RESTARTS   AGE
coredns-2955756869-lkjgk                0/1     Running   0          28s
coredns-2955756869-tps0t                0/1     Running   0          28s
kubernetes-dashboard-1724149408-97377   1/1     Running   3          5d
```

从上面的输出结果可以看到，CoreDNS 已经有 2 个实例在运行。

接下来，修改 kubelet 的配置文件/etc/kubernetes/kubelet，增加关于 DNS 的相关选项，如下所示：

```
KUBELET_ARGS="--cluster-dns=10.254.0.2 --cluster-domain=cluster.local"
```

查看相应的服务的状态：

```
[root@localhost ~]# kubectl get svc --namespace=kube-system
NAME                   CLUSTER-IP     EXTERNAL-IP   PORT(S)                  AGE
kube-dns                10.254.0.2      <none>      53/UDP,53/TCP,9153/TCP   48m
kubernetes-dashboard   10.254.175.41   <nodes>      9090:30963/TCP            5d
```

从上面的输出结果可知，kube-dns 服务的 ClusterIP 为 10.254.0.2，正是我们在配置文件中指定的 IP 地址。此外，该服务暴露的端口为 53/UDP、53/TCP 以及 9153/TCP。

接下来，验证所部署的 CoreDNS 是否可以正常将服务名解析为对应的 IP 地址。首先部署一个 Nginx 及其服务，其中 Deployment 的 YAML 代码如下：

```
apiVersion: extensions/v1beta1
kind: Deployment
metadata:
  name: my-nginx
spec:
  selector:
    matchLabels:
      run: my-nginx
  replicas: 2
  template:
    metadata:
      labels:
        run: my-nginx
    spec:
      containers:
      - name: my-nginx
        image: nginx
        ports:
        - containerPort: 80
```

服务的 YAML 代码如下：

```
apiVersion: v1
kind: Service
metadata:
  name: my-nginx
  labels:
    run: my-nginx
spec:
  ports:
  - port: 80
    protocol: TCP
  selector:
    run: my-nginx
```

创建一个 CentOS 的 Pod，验证 CoreDNS 是否能够成功解析名称，其 YAML 配置文件如下：

```
apiVersion: v1
kind: Pod
metadata:
  name: centos
  namespace: default
spec:
  containers:
  - image: centos
    command:
      - sleep
      - "3600"
    imagePullPolicy: IfNotPresent
    name: centoschao
  restartPolicy: Always
```

然后执行以下命令进入 CentOS 容器里面：

```
[root@localhost ~]# kubectl exec -it centos -- /bin/sh
sh-4.2#
```

通过 nslookup 命令验证是否可以解析 IP 地址：

```
sh-4.2# nslookup my-nginx
Server:         10.254.0.2
Address:        10.254.0.2#53

my-nginx
Name:   my-nginx.default.svc.cluster.local
Address: 10.254.29.191
```

从上面的输出结果可知，CoreDNS 已经成功地将服务名称解析为相应的 IP 地址。

注　　意

如果 CentOS 容器中没有安装 nslookup 命令，可以使用以下命令安装：

```
sh-4.2# yum -y install bind-utils
```

通过 curl 命令访问 my-nginx 服务，如下所示：

```
sh-4.2# curl my-nginx
<!DOCTYPE html>
<html>
<head>
<title>Welcome to nginx!</title>
```

```
<style>
    body {
        width: 35em;
        margin: 0 auto;
        font-family: Tahoma, Verdana, Arial, sans-serif;
    }
</style>
</head>
<body>
<h1>Welcome to nginx!</h1>
<p>If you see this page, the nginx web server is successfully installed and
working. Further configuration is required.</p>

<p>For online documentation and support please refer to
<a href="http://nginx.org/">nginx.org</a>.<br/>
Commercial support is available at
<a href="http://nginx.com/">nginx.com</a>.</p>

<p><em>Thank you for using nginx.</em></p>
</body>
</html>
```

第6章 存储管理

在 Kubernetes 集群中，服务的运行离不开将数据持久化地保存起来，这就涉及 Kubernetes 的存储系统了。存储系统的功能是将各种服务在运行过程中产生的数据长久地保存下来，即使容器被销毁，数据仍然存在。本章将对 Kubernetes 的存储系统进行详细介绍。

本章涉及的知识点有：

- *存储卷：主要介绍 Kubernetes 中存储卷的概念以及各种类型的存储卷的使用方法。*
- *持久化存储卷：介绍持久化存储卷的创建、回收以及动态供给。*

6.1 存 储 卷

存储卷（Volume）是 Kubernetes 持久化数据的最基本的功能单元。Kubernetes 的数据卷是 Docker 数据卷的扩展，Kubernetes 适配各种存储系统，包括本地存储 EmptyDir 和 HostPath，网络存储 NFS、GlusterFS 以及 PV/PVC 等。本节将详细介绍如何实现 Kubernetes 的存储。

6.1.1 什么是存储卷

我们通常讲，容器和 Pod 是短暂的，它们会被频繁地销毁和创建。容器被销毁时，保存在容器内部文件系统中的数据就会被清除。此外，Pod 中的多个容器经常需要共享文件，如果没有其他的机制，单纯依靠容器本身是无法实现的。正因为以上的原因，Kubernetes 专门提供了存储卷（Volume）来解决这些问题。

存储卷的生命周期独立于容器，Pod 中的容器可能被销毁和重建，但存储卷会被保留。本质上，Kubernetes 的存储卷是一个目录，这一点与 Docker 的卷类似。当存储卷被挂载到 Pod，Pod 中的所有容器都可以访问这个存储卷。Kubernetes 的存储卷也支持多种后端类型，目前为止，大约有 30 余种，主要包括 emptyDir、hostPath、GCE Persistent Disk、AWS Elastic BlockStore、NFS 以及 Ceph 等。

存储卷提供了对各种后端存储的抽象，容器在使用存储卷读写数据的时候，不需要关心数据到底是存放在本地节点的文件系统中，还是云硬盘上。对它来说，所有类型的存储卷都只是一个目录。

6.1.2 emptyDir 卷

emptyDir 卷是最基础的存储卷类型之一。简单地讲，一个 emptyDir 类型的存储卷就是宿主节点上面的一个空目录。

emptyDir 卷对于容器来说是持久的，也就是说，emptyDir 卷不会随着容器的销毁而销毁。但是 emptyDir 卷对于 Pod 来说，则不是持久的。当 Pod 从节点中删除时，其所拥有的 emptyDir 卷也会被删除，其中的数据也会丢失。也就是说，emptyDir 卷与 Pod 的生命周期是一致的。Pod 中的所有容器都可以共享卷，它们可以指定各自的挂载路径。下面通过例子来演示 emptyDir 卷的使用方法。首先创建一个名为 emptydir-demo.yaml 的配置文件，内容如下：

```
01 apiVersion: v1
02 kind: Pod
03 metadata:
04   name: emptydir-demo
05 spec:
06   containers:
07     - name: c1
08       image: nginx:latest
09       volumeMounts:
10       - mountPath: /messages
11         name: data
12       args:
13       - /bin/bash
14       - -c
15       - echo "Hello, world." > /messages/hello;sleep 3000
16     - name: c2
17       image: nginx:latest
18       volumeMounts:
19       - mountPath: /messages
20         name: data
21       args:
22       - /bin/bash
23       - -c
24       - cat /messages/hello;sleep 3000
25   volumes:
26     - name: data
27       emptyDir: {}
```

在上面的代码中，第 2 行指定资源类型为 Pod，该 Pod 将会挂载所创建的 emptyDir。第 4 行指定 Pod 的名称为 emptydir-demo。第 7~15 行定义了一个容器，该容器的名称为 c1。第 9~11 行定义容器内的挂载点，挂载路径为/messages，存储卷的名称为 data。第 13~15 行定义参数，该参数的功能是输出以下字符串：

```
Hello, world.
```

以上字符串将会被输出到/messages/hello 文件中，并且休眠 3 秒。

第 16~24 行定义了另外一个容器，其名称为 c2。c2 同样也将名称为 data 的 emptyDir 卷挂载在/messages 路径下。不过，该容器的参数指定通过 cat 命令将/messages/hello 文件的内容输出到标准输出。第 25 行开始定义 emptyDir 卷，其名称为 data，类型为 emptyDir。

通过上面介绍可知，上面的例子在名为 emptydir-demo 的 Pod 中定义了 2 个容器，其中一个容器的功能是输出一个字符串到 emptyDir 卷上的文件中，另外一个容器的功能是读取该文件。这两个容器同时挂载同一个 emptyDir 卷，实现了存储的共享。

然后使用以下命令创建 emptyDir 卷：

```
[root@localhost ~]# kubectl create -f emptydir-demo.yaml
pod "emptydir-demo" created
```

创建完成之后，查看所创建的 Pod 的状态，如下所示：

```
[root@localhost ~]# kubectl get po -o wide
NAME          READY   STATUS      RESTARTS     AGE  IP           NODE
emptydir-demo 2/2     Running     0            36m  172.16.10.3  192.168.1.122
…
```

从上面的输出结果可知，刚刚创建的 Pod 位于 192.168.1.122 节点上。通常情况下，存储卷在节点上的磁盘路径为：

```
/var/lib/kubelet/pods/<pod uuid>/volumes
```

其中 pod uuid 为 Kubernetes 分配给 Pod 本身的 UUID。这个 UUID 可以通过 kubectl get pod 命令获取，需要输出 JSON 格式的结果，如下所示：

```
[root@localhost ~]# kubectl get pod emptydir-demo -o json
{
   "apiVersion": "v1",
   "kind": "Pod",
   "metadata": {
      "creationTimestamp": "2019-03-29T21:29:37Z",
      "name": "emptydir-demo",
      "namespace": "default",
      "resourceVersion": "136078",
      "selfLink": "/api/v1/namespaces/default/pods/emptydir-demo",
      "uid": "c13c2c25-5269-11e9-a06b-000c2994a2b7"
   },
   "spec": {
      "containers": [
         {
            "args": [
               "/bin/bash",
```

```
                    "-c",
                    "echo \"Hello, world.\" \u003e /messages/hello;sleep 3000"
                ],
                "image": "nginx:latest",
                "imagePullPolicy": "Always",
                "name": "master",
                "resources": {},
                "terminationMessagePath": "/dev/termination-log",
                "volumeMounts": [
                    {
                        "mountPath": "/messages",
                        "name": "data"
                    }
                ]
            },
...
```

从上面的输出结果可知，emptydir-demo 的 UUID 为 c13c2c25-5269-11e9-a06b-000c2994a2b7。得知 UUID 之后，我们就可以登录到 IP 地址为 192.168.1.122 的节点，查看以下目录：

```
/var/lib/kubelet/pods/c13c2c25-5269-11e9-a06b-000c2994a2b7/volumes
```

目录内容如下：

```
[root@localhost ~]# ll /var/lib/kubelet/pods/c13c2c25-5269-11e9-a06b-000c2994a2b7/volumes
total 0
drwxr-xr-x  3   root    root    18 Mar 30 05:29     kubernetes.io~empty-dir
```

以上输出结果中的 kubernetes.io~empty-dir，即为 emptyDir 类型的存储卷所在的目录。在该目录中，保存着我们上面创建的名为 data 的 emptyDir 卷。在 data 目录中，会发现一个名为 hello 的文本文件，该文件的内容如下：

```
[root@localhost ~]# cat /var/lib/kubelet/pods/c13c2c25-5269-11e9-a06b-000c2994a2b7/volumes/kubernetes.io~empty-dir/data/hello
Hello, world.
```

上面文件的内容就是名为 c1 的容器的命令的输出结果。

而对于第 2 个容器的输出结果，我们可以在容器的日志中获取到，如下所示：

```
[root@localhost ~]# kubectl logs emptydir-demo c2
Hello, world.
```

在上面的命令中，emptydir-demo 为 Pod 的名称，c2 为 Pod 中的容器的名称。

接下来，我们再分别在两个容器里面验证一下，看看是否能够访问刚才定义的存储卷。

首先查看名为 c1 的容器，由于在配置文件中我们指定卷的挂载路径为/messages，因此命令如下：

```
[root@localhost ~]# kubectl exec -it emptydir-demo -c c1 -- ls -l /messages
total 4
-rw-r--r-- 1    root        root    14 Mar 29 22:19          hello
```

从上面的输出可以确认，在c1中的/messages目录中确实存在着名为hello的文件，其内容如下：

```
[root@localhost ~]# kubectl exec -it emptydir-demo -c slave -- cat
/messages/hello
Hello, world.
```

同样在c2中，也可以得到类似的结果，读者可以自行验证。

如果我们在节点中执行以下命令，在emptyDir卷所在的目录中创建一个新的文件：

```
[root@localhost ~]# echo "a message" > /var/lib/kubelet/pods/c13c2c25-
5269-11e9-a06b-000c2994a2b7/volumes/kubernetes.io~empty-dir/data/message
```

那这个文件也可以在容器中访问到，如下所示：

```
[root@localhost ~]# kubectl exec -it emptydir-demo -c slave -- cat
/messages/message
a message
```

6.1.3 hostPath 卷

hostPath类型的存储卷的作用是，将节点的文件系统中已经存在的目录直接共享给Pod的容器。在实际生产环境中，大部分应用都不会使用hostPath卷，因为这实际上增加了Pod与节点的耦合，限制了Pod的使用。但是，如果应用系统需要访问Kubernetes或Docker内部数据，例如配置文件和二进制库，则需要使用hostPath卷。

hostPath卷一般和DaemonSet搭配使用，用来操作主机文件，例如加载主机的容器日志目录，达到收集本主机所有日志的目的。下面的代码是一个hostPath的YAML配置文件：

```
apiVersion: v1
kind: Pod
metadata:
  name: hostpath-demo
spec:
  containers:
  - image: nginx
    name: test-container
    volumeMounts:
    - mountPath: /data
      name: test-volume
  volumes:
  - name: test-volume
    hostPath:
```

```
      # directory location on host
      path: /root/data
```

假设在节点的文件系统中存储着一个目录，其路径为：

```
/root/data
```

上面的配置文件就是使得名为 test-container 的容器直接挂在主机的/root/data 目录。

6.1.4 NFS 卷

NFS 即网络文件系统，它最早在 FreeBSD 中实现。NFS 的主要功能是通过局域网让不同的主机之间共享文件或者目录。NFS 是典型的客户机/服务器架构，其客户端通常为应用服务器，而服务器通常是连接大容量存储设备的主机。客户机通过挂载的方式，将 NFS 服务器共享出来的目录挂载到本地系统中。从客户机的角度看，NFS 服务器共享出来的目录，就好像是自己本地的磁盘分区或者目录一样，而实际上所有的文件系统都在服务器上面。

注　意
NFS 中的文件系统属于 NFS 服务器，而不属于客户机，这一点与 iSCSI 有着本质区别。

NFS 网络文件系统类似 Windows 系统中的网络共享和网络驱动器映射，也和 Linux 系统里的 Samba 服务类似。

在企业集群架构的工作场景中，NFS 网络文件系统一般被用来存储共享视频、图片、附件等静态资源文件。一般是把网站用户上传的文件都放在 NFS 中共享，例如，BBS 产品的图片、附件、头像。但是要注意的是，网站 BBS 程序不要放在 NFS 中共享，然后让前端所有的节点访问存储服务，这种不规范的用法在中小网站公司中应用频率很高。

Kubernetes 的存储卷支持 NFS 类型的远程存储系统，允许将一块现有的网络硬盘在同一个 Pod 内的容器间共享。

例如，当前局域网中存在着一个 NFS 服务器，其 IP 地址为 192.168.12.40，该服务器将一个本地的文件系统/data/dsk1，通过 NFS 协议共享给 Kubernetes 集群使用。

```
apiVersion: v1
kind: Pod
metadata:
  labels:
    name: nfsdemo
    role: master
  name: nfspathpod
spec:
  containers:
  - name: c1
    image: nginx
    volumeMounts:
```

```
        - name: nfs-storage
          mountPath: /nfs/
      volumes:
      - name: nfs-storage
        nfs:
          server: 192.168.1.40
          path: "/data/dsk1"
```

注 意

使用 NFS 时需要在节点上安装 NFS 文件系统相关组件，否则节点无法挂载 NFS 文件系统。

6.1.5 Secret 卷

在 Kubernetes 中，Secret 卷是用来保存小片敏感数据的存储卷，例如账号、密码或者秘钥等。对于这类数据，管理员必须妥善保管。当然，除了妥善保管之外，还必须能够非常方便地控制如何使用。而 Secret 卷正是出于这种目的而设计的存储卷，相比于直接将敏感数据配置在 Pod 的定义或者镜像中，Secret 卷提供了更加安全的 Base64 加密方法，防止数据泄露。用户可以创建自己的 Secret 卷，Kubernetes 系统也会有自己本身的 Secret 卷。

Pod 有两种方式使用 Secret 卷，首先，Secret 卷可以作为存储卷被一个或者多个容器挂载；其次，在拉取镜像文件的时候被 kubelet 引用。

Secret 卷的创建是独立于 Pod 的，以数据卷的形式挂载到 Pod 中，Secret 卷的数据将以文件的形式保存，文件中保存的是一个或者多个“键-值对”（Key-Value Pair），容器通过读取文件可以获取需要的数据。下面详细介绍如何使用 Secret 卷存储账号和密码数据。

创建 Secret 卷的命令如下：

```
    kubectl create secret name [--type=string] [--from-file=[key=]source]
[--from-literal=key1=value1]
```

用户可以通过 3 种方式为以上命令提供参数，分别是配置文件、目录或者字符串。当通过配置文件指定参数时，配置文件的基础文件名，即去掉路径和扩展名，将被作为“键-值对”的键，文件的内容将作为键值对的值。例如，如果配置文件的文件名为 username.txt，内容为 admin，则对应 Secret 卷中的“键-值对”为：

```
username=admin
```

如果用户通过目录创建 Secret 卷，则指定目录中的每个常规文件的基础文件名都将作为“键-值对”的键，其内容作该键对应的值。

如果直接在命令行中指定“键-值对”，则其语法如下：

```
--from-literal=key1=supersecret --from-literal=key2=topsecret
```

其中 key1 和 key2 为“键-值对”的键，supersecret 和 topsecret 分别为对应的值。

例如，用户在访问数据库的时候，需要使用用户名和密码，这些账户信息都要存储到 Secret 卷中，其操作步骤如下。

（1）准备两个文本文件，其文件名分别为 username.txt 和 password.txt，命令如下：

```
[root@localhost ~]# echo -n 'admin' > username.txt
[root@localhost ~]# echo -n 'd5eeff42' > password.txt
```

上面的命令中，-n 选项表示不输出最后的换行。大于号为重定向运算符，表示将 echo 命令的输出结果重定向到文件中。

（2）使用 kubectl create 命令创建 Secret 卷，如下所示：

```
[root@localhost ~]# kubectl create secret generic db-secret
--from-file=./username.txt --from-file=./password.txt
```

通过上面的命令可知，用户可以使用多个--from-file 命令来通过“键-值对”配置文件。

（3）查看 Secret 卷，命令如下：

```
[root@localhost ~]# kubectl get secrets
NAME            TYPE              DATA              AGE
db-secret   Opaque             2                9m
```

从上面的输出结果可知，所创建的 Secret 卷的名称为 db-secret，类型为 Opaque，即不透明的，包含 2 个“键-值对”。

（4）查看 Secret 卷的详细信息，命令如下：

```
[root@localhost ~]# kubectl describe secrets/db-secret
Name:           db-secret
Namespace:       default
Labels:         <none>
Annotations:     <none>

Type:   Opaque

Data
====
password.txt:   8 bytes
username.txt:   5 bytes
```

除了使用配置文件之外，以上操作可以通过命令行参数更加便捷地完成，相应的命令如下：

```
[root@localhost ~]# kubectl create secret generic db-secret
--from-literal=username='admin' --from-literal=password='d5eeff42'
```

当然了，作为一种 Kubernetes 资源，Secret 卷也可以通过 YAML 配置文件来创建。例如，上面的 Secret 卷的 YAML 配置文件内容如下：

```
apiVersion: v1
kind: Secret
metadata:
  name: db-secret
type: Opaque
data:
  username: admin
  password: d5eeff42
```

将上面的 YAML 配置文件保存为 db-secret.yaml，然后使用以下命令创建 Secret 卷：

```
[root@localhost ~]# kubectl create -f db-secret.yaml
```

接下来我们创建一个 Pod，用来挂载前面创建的 Secret 卷，并且使用里面存储的账号信息。该 Pod 的 YAML 配置文件的名为 test-secret.yaml，其内容如下：

```
apiVersion: v1
kind: Pod
metadata:
  labels:
    name: test-secret
    role: master
  name: test-secret
spec:
  containers:
  - name: test-secret
    image: nginx
    volumeMounts:
      - name: secret
        mountPath: /home/iron/secret
        readOnly: true
  volumes:
  - name: secret
   secret:
      secretName: db-secret
```

在上面的配置文件中，将前面定义的 db-secret 挂载到容器中。然后使用以下命令创建 Pod：

```
[root@localhost ~]# kubectl create -f test-secret.yaml
```

我们再验证一下能否从容器中访问到 db-secret 中存储的数据。执行以下命令，查看 db-secret 数据卷中的文件列表，如下所示：

```
[root@localhost ~]# kubectl exec -it test-secret -c test-secret -- ls -l
/home/iron/secret
total 0
```

```
    lrwxrwxrwx  1   root    root    19 Mar 30 21:20     password.txt
-> ..data/password.txt
    lrwxrwxrwx  1   root    root    19 Mar 30 21:20     username.txt
-> ..data/username.txt
```

从上面的输出结果可知，前面存储在 db-secret 中的两个文件已经被列出来了。接下来通过 cat 命令查看文件的内容：

```
    [root@localhost ~]# kubectl exec -it test-secret -c test-secret -- cat
/home/iron/secret/username.txt
    admin[root@localhost ~]#
    [root@localhost ~]# kubectl exec -it test-secret -c test-secret -- cat
/home/iron/secret/password.txt
    d5eeff42[root@localhost ~]#
```

6.1.6 iSCSI 卷

iSCSI 卷允许将现有的 iSCSI 磁盘挂载到用户的 Pod 中，与 emptyDir 不同的是，删除 Pod 时 emptyDir 会被删除，但 iSCSI 卷只是被卸载，内容则会被保留。

下面的代码为一个挂载 iSCSI 卷的 Pod 的 YAML 配置：

```
apiVersion: v1
kind: Pod
metadata:
  name: iscsipd
spec:
  containers:
  - name: iscsipd-rw
    image: kubernetes/pause
    volumeMounts:
    - mountPath: "/mnt/iscsipd"
      name: iscsipd-rw
  volumes:
  - name: iscsipd-rw
    iscsi:
      targetPortal: 10.0.2.15:3260
      portals: ['10.0.2.16:3260', '10.0.2.17:3260']
      iqn: iqn.2001-04.com.example:storage.kube.sys1.xyz
      lun: 0
      fsType: ext4
      readOnly: true
```

关于更加详细的 iSCSI 存储卷的使用方法，请参考相关的技术手册，这里不再具体介绍。

6.2 持久化存储卷

上一节介绍的存储卷已经为 Kubernetes 的数据持久化提供了很好的解决方案。但是存储卷在可管理性方面有着比较大的缺陷。于是 Kubernetes 又提出了持久化存储卷来解决这个问题。本节将详细介绍持久化存储卷的使用方法。

6.2.1 什么是持久化存储卷

通过 6.1 节的学习，我们已经掌握了许多存储卷的使用方法。从前面的介绍可以得知，用户在使用存储卷的时候，必须首先掌握关于存储卷的各种细节。例如，在使用 NFS 存储卷的时候，用户需要知道 NFS 服务器的相关信息，例如 IP 地址、共享路径以及账号信息等。但是，在实际开发和运维过程中，Pod 通常由开发人员来管理，而存储卷通常由维护人员来管理。这会导致开发人员和维护人员的工作边界不清晰，由此产生许多管理上的问题。如果系统规模较小或者对于开发环境，这样的情况还可以接受。但当集群规模变大，特别是对于生成环境，考虑到效率和安全性，这就成了必须解决的问题。

持久化存储卷（PersistentVolume）正是为了解决这个管理上的问题而提出的。持久化存储卷和节点一样，都是 Kubernetes 集群中的一种资源，也同样存在着独立的生命周期。但是，持久化存储卷和存储卷不同之处在于：持久化存储卷屏蔽了底层存储的实现细节，方便普通的用户使用，同时能方便管理员管理。

Kubernetes 的持久化存储卷支持非常多的存储类型，例如 gcePersistentDisk、AWSElasticBlockStore、AzureFile、AzureDisk、FC 存储、NFS 网络文件系统、iSCSI 以及 GlusterFS 等。每种存储类型都有各自的特点，在使用时需要根据它们各自的参数进行设置。

6.2.2 持久化存储卷请求

持久化存储卷请求（PersistentVolumeClaim）描述了普通用户对持久化存储卷的需求，由普通用户创建和维护。当普通用户需要为 Pod 分配存储资源时，就创建一个持久化存储卷请求，指明了自己所需要的存储资源的容量和访问方式等信息，Kubernetes 就会自动在集群中查找并提供符合要求的持久化存储卷，提供给 Pod 使用。

从上面的描述可以得知，在使用持久化存储卷的时候，普通用户只关心自己有多少存储资源以及如何使用存储资源，并不需要关心这些存储资源从何而来，具体的实现方式是什么；而对于管理员来说，只需要根据用户提供的声明的需求，来创建并分配持久化存储卷就可以了。

6.2.3 持久化存储卷生命周期

持久化存储卷及其请求都是 Kubernetes 的资源，有着自己独立的生命周期。在整个生命周期中，持久化存储卷及其请求相互作用，如图 6-1 所示。

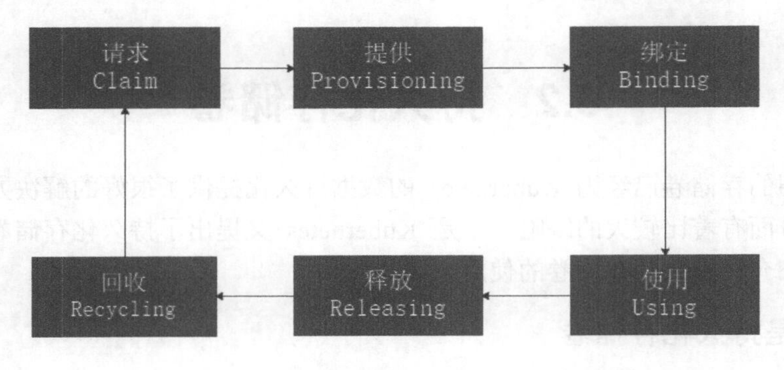

图 6-1　持久化存储卷生命周期

1. 提供（Provisioning）

管理员可以通过两种方式提供持久化存储卷，分别为静态和动态：

- 静态提供（Static）：管理员在 Kubernetes 集群中手工创建持久化存储卷，根据用户请求提供给 Pod 使用。
- 动态提供（Dynamic）：当目前集群中没有符合用户请求的持久化存储卷的时候，集群会尝试根据用户请求动态生成存储卷。

2. 请求（Claiming）

普通用户根据自己的业务需求，在集群中创建存储资源请求，在请求中描述自己对于存储容器以及访问模式。

3. 绑定（Binding）

当用户在集群中发起一个新的存储请求时，Kubernetes 的控制器会试图根据请求中的存储大小以及访问模式等条件，查找最合适的存储卷并建立绑定关系。这里“最合适的”意思是存储卷一定满足请求的最低要求，但是也可能比请求的要多。例如，用户请求 10GB 存储资源，但当前集群中最小的持久化存储卷是 15GB，那么这个存储卷也会被分配给用户。

注　　意
一个持久化存储卷只能绑定给一个用户请求。

4. 使用（Using）

Pod 把持久化存储卷当作是一个普通的存储卷来使用，与前面介绍的各种存储卷的使用方法并没有明显的区别。关于持久化存储卷的详细使用方法，将在随后介绍。

5. 释放（Releasing）

当用户使用完持久化存储卷后，就可以把发起的请求删除，绑定在该请求上面的持久化存储卷会变成 released 状态，并准备被回收。

6. 回收（Recycling）

Kubernetes会根据回收策略回收处于released状态的持久化存储卷，目前有3种回收策略：

- Retained：持久化存储会保持原有数据并允许用户手动回收数据。
- Recycled：Kubernetes 会彻底删除持久化存储卷中的数据，并允许该卷被绑定到其他用户请求，数据卷不会被删除。
- Deleted：删除数据并删除存储卷。

6.2.4 持久化存储卷静态绑定

图 6-2 描述了容器、持久化存储卷请求以及持久化存储卷之间的关系。从图中可以得知，存储卷的静态绑定可以分为 3 个部分，其中容器应用调用存储卷请求，存储卷的请求描述了容器对于存储资源的需求，存储卷则是存储资源的提供者。

图 6-2 容器应用、存储卷请求以及存储卷之间的关系

下面分别按照这 3 个部分介绍持久存储卷的静态绑定。

（1）创建应用，YAML 配置的代码如下：

```
01 apiVersion: extensions/v1beta1
02 kind: Deployment
03 metadata:
04   name: nginx-deployment
05 spec:
06   replicas: 2
07   selector:
08     matchLabels:
09       app: nginx
10   template:
11     metadata:
12       labels:
13         app: nginx
14     spec:
15       containers:
16       - name: nginx
17         image: nginx
18         volumeMounts:
```

```
19          - name: wwwroot
20            mountPath: /usr/share/nginx/html
21          ports:
22          - containerPort: 80
23        volumes:
24        - name: wwwroot
25          persistentVolumeClaim:
26            claimName: my-pvc
```

其中，第 25~26 行定义了持久化存储请求的名称为 my-pvc。

使用以下命令创建应用：

```
[root@localhost ~]# kubectl apply -f nginx-deployment.yaml
```

命令中 nginx-deployment.yaml 为上面的 YAML 配置文件的文件名。

（2）定义持久化存储请求，YAML 配置文件的内容如下：

```
01 apiVersion: v1
02 kind: PersistentVolumeClaim
03 metadata:
04   name: my-pvc
05 spec:
06   accessModes:
07     - ReadWriteMany
08   resources:
09     requests:
10       storage: 1Gi
```

其中，第 2 行指定资源的类型为 PersistentVolumeClaim。第 4 行定义持久化存储卷的卷名为 my-pvc。第 7 行指定存储卷的访问模式为 ReadWriteMany。第 10 行指定卷的内容为 1GB。

使用以下命令创建资源请求，其中 my-pvc.yaml 为上面 YAML 文件的文件名：

```
[root@localhost ~]# kubectl apply -f my-pvc.yaml
```

（3）定义持久化存储卷，YAML 的配置文件如下：

```
01 apiVersion: v1
02 kind: PersistentVolume
03 metadata:
04     name: my-pv
05 spec:
06     capacity:
07       storage: 2Gi
08     accessModes:
09       - ReadWriteMany
10     persistentVolumeReclaimPolicy: Recycle
11     nfs:
```

```
12          path: /data/dsk1
13          server: 192.168.1.141
```

第 8 行定义了存储卷的访问模式为 ReadWriteMany，这个访问模式必须与前面的 my-pvc.yaml 对存储卷需求的定义完全相同，否则，会出现无法匹配成功的错误。第 10 行定义了卷的回收策略。

创建持久化存储卷的命令如下：

```
[root@localhost ~]# kubectl apply -f pv.yaml
```

创建完成之后，通过以下命令查看持久化存储卷的状态，如下所示：

```
[root@localhost ~]# kubectl get pv
NAME      CAPACITY   ACCESSMODES   RECLAIMPOLICY   STATUS    CLAIM
REASON    AGE
my-pv     2Gi        RWX           Recycle         Bound     default/my-pvc
56m
```

在上面的输出中，CAPACITY 为存储卷的容量，这个卷的容量为 2GB。ACCESSMODES 为卷的访问模式，RWX 表示可读、可写和可执行。RECLAIMPOLICY 为回收策略。STATUS 为卷的状态，其中 Bound 表示该卷已经与请求绑定。CLAIM 为存储请求的名称，default/my-pvc 中的 default 为命名空间，my-pvc 正是我们前面定义的请求的名称。

查看持久化存储请求的状态，如下所示：

```
[root@localhost ~]# kubectl get pvc
NAME          STATUS    VOLUME    CAPACITY   ACCESSMODES        AGE
pvc/my-pvc    Bound     my-pv     2Gi        RWX                52m
```

STATUS 为请求的状态，Bound 表示该请求已经与存储卷绑定，VOLUME 为存储卷的卷名，my-pv 正是我们上面定义的持久化存储卷。CAPACITY 为存储容量，其值为 2GB。可以发现，在前面的定义中，我们指定了所需要的容量为 1GB，但是当前集群中并没有完全符合该请求的存储卷，除了一个容量为 2GB 的 my-pv，所以 Kubernetes 就把 my-pv 提供给了 my-pvc，这也说明了 Kubernetes 在将存储请求和存储卷绑定时，采用的是最低匹配原则。ACCESSMODES 为访问模式，这个访问模式与前面定义的存储卷的访问模式一致。

最后再查看一下 Pod 的详细信息，如下所示：

```
[root@localhost ~]# kubectl describe pod nginx-deployment-3793384989-7jq4l
Name:           nginx-deployment-3793384989-7jq4l
Namespace:      default
Node:           192.168.1.122/192.168.1.122
Start Time:     Mon, 01 Apr 2019 00:01:51 +0800
Labels:         app=nginx
                pod-template-hash=3793384989
Status:         Running
IP:             172.16.10.5
```

```
    Controllers:    ReplicaSet/nginx-deployment-3793384989
    Containers:
      nginx:
        Container ID:
docker://1fa15cbfdb6f9f0e02ab3b9fbd3c0864af4bdd30719d4636de76332b4ebe3812
        Image:          nginx
        Image ID:
docker-pullable://docker.io/nginx@sha256:8569b2ded35ade8dafe00587edd6962d05db
5084aaed3c5485d8ce168790aa07
        Port:           80/TCP
        State:              Running
          Started:          Mon, 01 Apr 2019 00:01:58 +0800
        Ready:            True
        Restart Count:      0
        Volume Mounts:
          /usr/share/nginx/html from wwwroot (rw)
        Environment Variables:      <none>
    Conditions:
      Type          Status
      Initialized   True
      Ready         True
      PodScheduled  True
    Volumes:
      wwwroot:
        Type:       PersistentVolumeClaim (a reference to a PersistentVolumeClaim
in the same namespace)
        ClaimName:  my-pvc
        ReadOnly:   false
    QoS Class:      BestEffort
    Tolerations:    <none>
    No events.
```

从上面的输出结果可知，该 Pod 的卷的类型为 PersistentVolumeClaim，其卷名为 my-pvc。

6.2.5　持久存储卷动态绑定

与静态持久存储卷绑定相比，动态持久存储卷绑定不需要预先创建存储卷，而是通过持久存储卷控制器动态调度，根据用户的存储资源请求，寻找 StorageClass 定义的、符合要求的底层存储来分配资源。

动态卷供给是 Kubernetes 独有的功能，这一功能允许按需创建存储卷。在此之前，集群管理员需要事先在集群外由存储提供者或者云提供商创建存储卷，成功之后再创建持久存储卷对象，才能够在 Kubernetes 中使用。动态卷供给能让集群管理员不必进行预先创建存储卷，而是随着用户需求进行创建。

Kubernetes 的持久存储卷绑定采用插件的形式提供。Kubernetes 官方内置了非常多的存储供应商的驱动程序（Provisioner），用户可以根据自己的需要来自由选择。

当然，除了内置的驱动程序之外，用户还可以选择另外的后端存储驱动程序。例如 NFS 或者 iSCSI 等。下面以 NFS 作为后端存储卷为例，来介绍动态绑定的方法。

在本例中，我们假定当前网络中已经存在着一个 NFS 服务器，该服务器的 IP 地址为 192.168.1.141，共享出来的目录为/data/dsk1。

（1）创建 Deployment。通过 git 命令克隆 Kubernetes 提供的外部存储卷驱动程序代码，如下所示：

```
    [root@localhost ~]# git clone
https://github.com/kubernetes-incubator/external-storage
```

克隆完成之后，进入 external-storage 目录下的 nfs-client 目录，命令如下：

```
    [root@localhost ~]# cd external-storage/
    [root@localhost external-storage]# cd nfs-client
```

然后修改 deploy 目录中的 deployment.yaml 配置文件，代码如下：

```
01 kind: Deployment
02 apiVersion: extensions/v1beta1
03 metadata:
04   name: nfs-client-provisioner
05 spec:
06   replicas: 1
07   strategy:
08     type: Recreate
09   template:
10     metadata:
11       labels:
12         app: nfs-client-provisioner
13     spec:
14       serviceAccountName: nfs-client-provisioner
15       containers:
16         - name: nfs-client-provisioner
17           image: quay.io/external_storage/nfs-client-provisioner:latest
18           volumeMounts:
19             - name: nfs-client-root
20               mountPath: /persistentvolumes
21           env:
22             - name: PROVISIONER_NAME
23               value: fuseim.pri/ifs
24             - name: NFS_SERVER
25               value: 192.168.1.141
```

```
26             - name: NFS_PATH
27               value: /data/dsk1
28       volumes:
29         - name: nfs-client-root
30           nfs:
31             server: 192.168.1.141
32             path: /data/dsk1
```

在上面的代码中，用户需要修改的地方主要是 NFS 服务器的 IP 地址和共享目录的路径，在本例中为第 25、27、31 和 32 行。其余的保留默认值即可。修改完成之后，使用以下命令创建 Deployment：

```
[root@localhost ~]# kubectl apply -f deploy/deployment.yaml
```

（2）创建 StorageClass。修改 StorageClass 配置文件，即 deploy 目录中的 class.yaml，其代码如下：

```
apiVersion: storage.k8s.io/v1
kind: StorageClass
metadata:
  name: managed-nfs-storage
provisioner: fuseim.pri/ifs # or choose another name, must match deployment's env PROVISIONER_NAME'
parameters:
  archiveOnDelete: "false"
```

然后使用以下命令创建 StorageClass：

```
[root@localhost ~]# kubectl apply -f deploy/class.yaml
```

使用以下命令查看刚刚创建的 StorageClass 的状态，如下所示：

```
[root@localhost ~]# kubectl get storageclass
NAME                          PROVISIONER          AGE
managed-nfs-storage           fuseim.pri/ifs       97m
```

（3）授权。用户需要执行以下命令进行授权：

```
[root@localhost ~]# kubectl create -f deploy/objects/serviceaccount.yaml
[root@localhost ~]# kubectl create -f deploy/objects/clusterrole.yaml
[root@localhost ~]# kubectl create -f deploy/objects/clusterrolebinding.yaml
[root@localhost ~]# kubectl patch deployment nfs-client-provisioner -p '{"spec":{"template":{"spec":{"serviceAccount":"nfs-client-provisioner"}}}}'
```

（4）创建持久化存储卷请求。Kubernetes 已经为用户提供了一个创建测试持久化存储卷请求的配置文件，其文件名为 deploy/test-claim.yaml，代码如下：

```
01 kind: PersistentVolumeClaim
02 apiVersion: v1
```

```
03 metadata:
04   name: test-claim
05   annotations:
06     volume.beta.kubernetes.io/storage-class: "managed-nfs-storage"
07 spec:
08   accessModes:
09     - ReadWriteMany
10   resources:
11     requests:
12       storage: 100Mi
```

上面代码的第 1 行指定要创建的资源类型为 PersistentVolumeClaim。第 6 行通过 volume.beta.kubernetes.io/storage-class 注解，定义对应的 StorageClass 为 managed-nfs-storage。第 12 行指定请求的存储空间为 100MB。

创建请求的命令如下：

```
[root@localhost ~]# kubectl apply -f deploy/test-claim.yaml
```

创建完成后，使用以下命令查看其状态：

```
[root@localhost ~]# kubectl get pvc
NAME      STATUS   VOLUME                                     CAPACITY    ACCESS
MODES    STORAGECLASS          AGE
test-claim  Bound    pvc-d2f8b670-55a2-11e9-94e5-000c29ecaa4e   100Mi
    RWX              managed-nfs-storage     52m
```

可以发现，存储卷请求的状态已经变成 Bound，其存储容量为 100Mi，其 StorageClass 为 managed-nfs-storage。

查看存储卷是否被自动创建，命令如下：

```
[root@localhost ~]# kubectl get pv
NAME                                       CAPACITY    ACCESS MODES    RECLAIM
POLICY       STATUS       CLAIM              STORAGECLASS          REASON      AGE
pvc-d2f8b670-55a2-11e9-94e5-000c29ecaa4e   100Mi       RWX
Delete                    Bound   default/test-claim      managed-nfs-storage
            18h
```

从上面的输出结果可知，Kubernetes 根据用户请求，自动创建了一个持久化存储卷。

（5）创建测试 Pod。修改 deploy 目录中的 test-pod.yaml 文件，代码如下：

```
01 kind: Pod
02 apiVersion: v1
03 metadata:
04   name: test-pod
05 spec:
06   containers:
```

```
07    - name: test-pod
08      image: nginx
09      command:
10        - "/bin/sh"
11      args:
12        - "-c"
13        - "touch /mnt/SUCCESS && exit 0 || exit 1"
14      volumeMounts:
15        - name: nfs-pvc
16          mountPath: "/mnt"
17    restartPolicy: "Never"
18    volumes:
19      - name: nfs-pvc
20        persistentVolumeClaim:
21          claimName: test-claim
```

在上面的代码中，第 13 行在容器创建完成之后，通过 touch 命令在/mnt 目录中创建一个名为 SUCCESS 的文件。第 14~16 行定义存储卷的挂载，将名为 nfs-pvc 的存储卷挂载到/mnt 上面。第 18~21 行定义存储卷，其中存储卷请求引用前面定义的 test-claim。

创建 Pod 的命令如下：

```
[root@localhost ~]# kubectl apply -f deploy/test-pod.yaml
```

查看 Pod 状态，如下所示：

```
[root@localhost ~]# kubectl get po
NAME                READY    STATUS       RESTARTS    AGE
…
test-pod            0/1      Completed    0           3m16s
…
```

从上面的输出可知，test-pod 已经处于完成状态。然后进入 NFS 服务器中的共享目录，查看是否 SUCCESS 文件：

```
[root@localhost ~]# ll
/data/dsk1/default-test-claim-pvc-d2f8b670-55a2-11e9-94e5-000c29ecaa4e/
total 0
-rw-r--r--    1    nfsnobody    nfsnobody   0   Apr  3 08:58
SUCCESS
```

从上面的输出结果可知，SUCCESS 文件已经被成功创建，这意味着 Pod 已经可以正常使用存储卷来存储数据了。同时，在整个过程中，我们并没有人工创建存储卷，只是创建了一个存储卷请求，Kubernetes 会自动根据请求创建持久化存储卷。

6.2.6 回收

用户可以通过删除持久化存储卷请求来达到回收存储资源的目的。存储资源请求被删除之后，持久化存储卷将变成 released 状态。由于还保留着之前的数据，这些数据需要根据不同的策略来处理，否则这些存储资源无法被其他存储资源请求使用。

对于持久化存储卷来说，用户可以指定 3 种回收策略。下面分别介绍这些回收策略的使用方法。

1. 保留（Retain）

在该种回收策略下，Kubernetes 允许用户手工回收存储资源，即当存储卷请求被删除后，存储卷将仍然存在，只是状态将会转换为已释放状态。对于其他的存储卷请求来说，处于已释放状态的是不可用的，因为以前的数据仍然保留在数据卷中。

我们以前面的静态绑定的持久化存储卷为例，来说明在保留策略情况下，删除 Pod 以及存储资源请求时数据状态的变化。

首先执行以下命令，在容器内部查看存储卷里面的数据，如下所示：

```
[root@localhost ~]# kubectl exec -it nginx-deployment-3793384989-p4jgf -c nginx -- ls -l /usr/share/nginx/html
total 0
-rw-r--r--  1  nobody    nogroup    0  Apr  2 23:53   a.txt
-rwxrwxrwx  1  root      root       0  Mar 30 12:12   test.txt
```

可以发现，在上面的存储卷中，一共有 2 个文件。

执行以下命令删除 Pod：

```
[root@localhost ~]# kubectl delete -f nginx-deployment.yaml
deployment "nginx-deployment" deleted
```

查看存储卷请求是否依然存在，如下所示：

```
[root@localhost ~]# kubectl get pvc
NAME         STATUS     VOLUME     CAPACITY    ACCESSMODES     AGE
my-pvc       Bound      my-pv      2Gi         RWX             14m
```

可以得知，尽管 Pod 被删除，但是存储卷请求依然存在。

继续删除存储卷请求，命令如下：

```
[root@localhost ~]# kubectl delete -f my-pvc.yaml
persistentvolumeclaim "my-pvc" deleted
```

查看存储卷的状态，如下所示：

```
[root@localhost ~]# kubectl get pv
NAME   CAPACITY  ACCESSMODES RECLAIMPOLICY   STATUS   CLAIM REASON      AGE
my-pv  2Gi       RWX             Retain       Released  default/my-pvc  18m
```

可以发现存储卷依然存在，只是 STATUS 的值由 Bound 转变为 Released。

登录到 NFS 服务器，查看存储卷保存在服务器上面的数据是否被删除，如下所示：

```
[root@localhost ~]# ll /data/dsk1/
total 0
-rw-r--r--      1    nfsnobody    nfsnobody    0    Apr  3 07:53    a.txt
-rwxrwxrwx      1    root         root         0    Mar 30 20:12    test.txt
```

从上面的输出结果可知，在存储卷请求被删除的情况下，存储卷的状态转换为释放状态，但是 Pod 存储 NFS 服务器上的文件依然存在。

继续删除持久化存储卷，命令如下：

```
[root@localhost ~]# kubectl delete -f pv.yaml
persistentvolume "my-pv" deleted
```

删除完成之后，再次在 NFS 服务器上面查询 Pod 存储的文件是否依然存在：

```
[root@localhost ~]# ll /data/dsk1/
total 0
-rw-r--r--      1    nfsnobody    nfsnobody    0    Apr  3 07:53    a.txt
-rwxrwxrwx      1    root         root         0    Mar 30 20:12    test.txt
```

可以发现，尽管持久化存储卷也被删除，但是存储在外部存储上面的文件，并没有随着存储卷的删除而被删除。这意味着在保留策略下，用户数据并不会随着相关资源的删除而被删除，其生命周期是独立的。

在数据确实不再需要的情况下，为了释放存储空间，用户需要手工在 NFS 服务器上面将文件删除。

在上面操作中，当 Pod 被删除后，存储卷请求是可以重用的，即如果在集群中，重新创建一个 Pod，可以直接将刚才的存储卷请求挂载。但是，当存储卷请求被删除，持久化存储卷的状态转换为释放状态时，即使当前集群中存在着符合条件的请求，该存储卷也不会被绑定。在存储卷被删除的时候，如果用户想要继续访问 NFS 服务器上的文件，可以重新创建存储卷，然后重新绑定请求。

2. 循环（Recycle）

如果持久化存储采用了循环回收策略，则在删除该卷时，存储在卷中的数据将被删除，使得存储卷可以与新的请求绑定。该策略将会在后面的版本中删除，建议使用动态绑定的方式来代替该策略。

3. 删除（Delete）

对于支持删除回收策略的存储卷插件，从集群中删除持久化存储卷的时候，也会从相关的外部设施中删除存储资产。

对于动态绑定来说，只支持删除策略。也就是说，如果采用动态绑定，在删除存储卷的时候，存储在外部存储，如 NFS 上面的所有数据都将被删除。

第7章 Kubernetes软件包管理

在操作系统当中，用户通常通过软件包管理工具来安装或者卸载软件，例如RPM、YUM以及Apt等。通过这些软件包管理工具，可以非常方便地对当前系统中的软件包进行管理。Kubernetes也提供了相应的软件包管理功能，即Helm。本章将对Helm的使用方法进行介绍。

本章涉及的知识点有：

- Helm：主要介绍Helm组件及相关术语。
- 安装Helm：介绍Helm客户端、服务器端以及客户端授权等操作。
- Chart：介绍Chart结构。
- Helm使用方法：主要介绍如何通过Helm对软件包进行管理。

7.1 Helm

Helm是Kubernetes生态系统中的一个软件包管理工具。本节将介绍Helm中的相关概念和基本工作原理，并通过一个具体的示例学习如何使用Helm打包、分发、安装、升级及回退Kubernetes应用。

7.1.1 Helm相关概念

Kubernetes是一个提供基于容器的应用集群管理解决方案，Kubernetes为容器化应用提供了部署运行、资源调度、服务发现和动态伸缩等一系列完整功能。

Kubernetes的核心设计理念是，用户定义要部署的应用程序的规则，而Kubernetes则负责按照定义的规则部署并运行应用程序。如果应用程序出现问题导致偏离了定义的规格，Kubernetes负责对其进行修正。例如，用户定义的应用规则要求部署两个Pod，如果在运行过程中，其中一个Pod异常终止了，Kubernetes会检查到并重新启动一个新的Pod实例。

用户通过使用Kubernetes API对象来描述应用程序规则，包括Pod、Service、Volume、Namespace、ReplicaSet、Deployment以及Job等。通常情况下，管理员在定义这些资源对象的时候，需要编辑一系列的YAML配置文件，然后通过Kubernetes命令行工具kubectl调用Kubernetes API进行部署。

以一个典型的三层应用 Wordpress 为例，该应用程序就涉及多个 Kubernetes API 对象，而要描述这些 Kubernetes API 对象，就可能要同时维护多个 YAML 文件。

因此，在进行 Kubernetes 软件部署时，管理员面临下述几个问题：

- 如何管理、编辑和更新这些分散的 Kubernetes 应用配置文件。
- 如何把一套相关的配置文件作为一个应用进行管理。
- 如何分发和重用 Kubernetes 的应用配置。

Helm 的出现就是为了很好地解决上面这些问题。

Helm 是一个用于 Kubernetes 应用的软件包管理工具，主要用来管理 Charts。有点类似于 Ubuntu 中的 APT 或 CentOS 中的 YUM。

对于应用发布者而言，可以通过 Helm 打包应用、管理应用依赖关系、管理应用版本，并发布应用到软件仓库。

对于使用者而言，使用 Helm 后就不用再编写复杂的 YAML 应用部署文件，可以以简单的方式在 Kubernetes 上查找、安装、升级、回滚、卸载应用程序。

简单地讲，Helm 是一个命令行的客户端工具，主要用于 Kubernetes 应用程序的创建、打包、发布以及创建和管理本地和远程的软件仓库。

7.1.2 Tiller

Tiller 是 Helm 的服务端，部署在 Kubernetes 集群中。Tiller 用于接收 Helm 的请求，并根据 Chart 生成 Kubernetes 的部署文件，即后面将要介绍的 Release，然后提交给 Kubernetes 创建应用。Tiller 还提供了 Release 的升级、删除、回滚等一系列功能。

7.1.3 Chart

Chart 即 Helm 所管理的软件包，采用 TAR 格式，类似于 APT 的 DEB 包或者 YUM 的 RPM 包，其包含了一组定义 Kubernetes 资源相关的 YAML 配置文件，可以在部署应用的时候自定义应用程序的一些元数据，以便于应用程序的分发。

7.1.4 Repository

Helm 的软件存储库称为 Repository。Repository 本质上是一个 Web 服务器，该服务器保存了一系列的 Chart 软件包以供用户下载，并且提供了一个该 Repository 所包含的 Chart 包的清单文件以供查询。Helm 可以同时管理多个不同的 Repository。

7.1.5 Release

使用 helm install 命令在 Kubernetes 集群中部署的 Chart 称为 Release。Release 实际上是 Helm 为某个 Chart 创建的实例。

7.2 安装 Helm

Helm 的安装方式很多，对于初学者来说，二进制软件包的方式最为简单。本节将介绍如何采用二进制的方式安装。更多安装方法可以参考 Helm 的官方帮助文档。

7.2.1 安装客户端

用户可以通过两种方式来安装 Helm 客户端，其中一种方式是通过官方提供的脚本一键式安装，另外一种方式是手动下载二进制文件。下面分别介绍这两种安装方式。

（1）首先介绍通过脚本一键式安装 Helm 客户端。Helm 在其官方的 GitHub 上面提供了一个名为 get 的 Shell 脚本文件，其网址如下：

https://github.com/helm/helm/blob/master/scripts/get

RAW 格式代码的网址如下：

https://raw.githubusercontent.com/helm/helm/master/scripts/get

用户可以通过以下命令将其代码下载到本地：

```
[root@localhost ~]# curl https://raw.githubusercontent.com/helm/helm/
master/scripts/get > get_helm.sh
```

其中 curl 命令是一个非常强大的 HTTP 客户端命令，大于号为 Shell 的重定向运算符，其功能是将 curl 命令的输出结果保存为 get_helm.sh 文件。下载完成后，修改 get_helm.sh 文件的权限，使其可以执行，命令如下：

```
[root@localhost ~]# chmod 700 get_helm.sh
```

最后执行该脚本文件，安装 Helm 客户端：

```
[root@localhost ~]# ./get_helm.sh
```

（2）接下来介绍通过手动下载预编译的二进制文件安装 Helm 客户端。Helm 二进制包的网址为：

https://github.com/helm/helm/releases/latest

Helm 为多种平台发布了预先编译好的二进制文件，如图 7-1 所示。

用户可以根据自己的实际环境下载所需要的版本。在本例中，我们采用的是 64 位的 CentOS，所以所需要的软件包的网址如下：

https://storage.googleapis.com/kubernetes-helm/helm-v3.4.2-linux-amd64.tar.gz

找到所需要的软件包之后，使用以下命令将其下载到本地：

```
[root@localhost ~]# wget https://storage.googleapis.com/kubernetes-helm/
helm-v3.4.2-linux-amd64.tar.gz
```

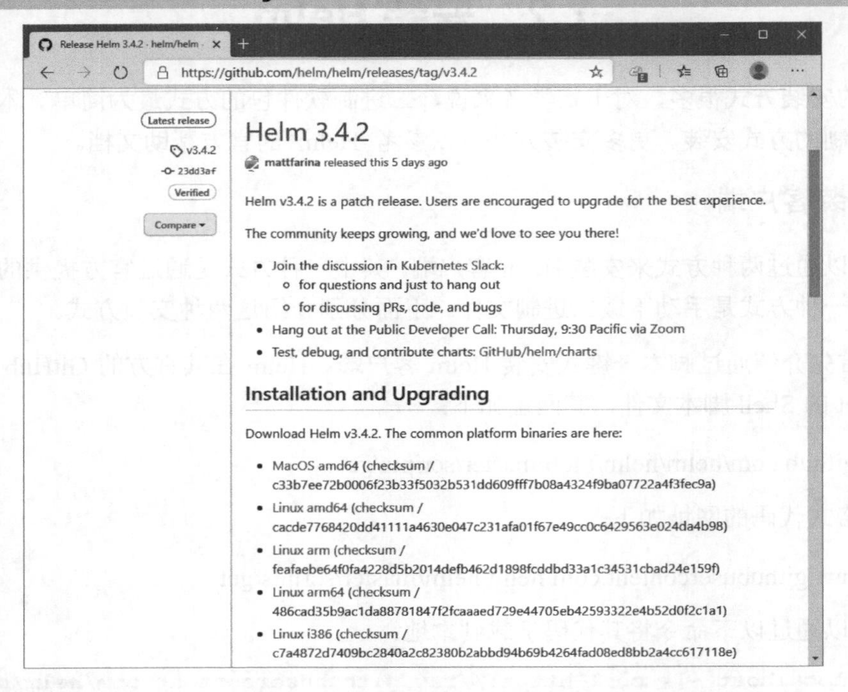

图 7-1　Helm 的二进制软件包

然后使用以下命令将其解压：

```
[root@localhost ~]# tar -zxvf helm-v3.4.2-linux-amd64.tar.gz
```

解压得到的目录名为 linux-amd64，查看其内容，如下所示：

```
[root@localhost ~]# ll linux-amd64/
total 72332
-rwxr-xr-x    1   root    root    37161248    Mar 22 02:44    helm
-rw-r--r--    1   root    root    11343       Mar 22 02:45    LICENSE
-rw-r--r--    1   root    root    3204        Mar 22 02:45    README.md
-rwxr-xr-x    1   root    root    36886016    Mar 22 02:45    tiller
```

在上面的文件列表中，helm 即我们所需要的客户端，tiller 为 Helm 的服务器端。

为了便于使用，可以将 helm 文件复制到系统目录/usr/local/bin 目录中，命令如下：

```
[root@localhost ~]# cp linux-amd64/helm /usr/local/bin/
```

7.2.2　安装服务端

Helm 的服务端 Tiller 是以 Deployment 方式部署在 Kubernetes 集群中的。通常情况下，管理员只需使用以下指令便可简单地完成安装：

```
[root@localhost ~]# helm init
```

但是由于在国内无法访问到Helm默认的服务器去下载镜像，故而不能直接使用以上命令。Helm 命令提供了一个-i 选项，用来指定自己的镜像服务器。

首先执行以下命令创建 helm 客户端相关配置文件以及$HELM_HOME 环境变量：

```
    [root@localhost ~]# helm init --client-only --stable-repo-url
https://aliacs-app-catalog.oss-cn-hangzhou.aliyuncs.com/charts/
    Creating /root/.helm
    Creating /root/.helm/repository
    Creating /root/.helm/repository/cache
    Creating /root/.helm/repository/local
    Creating /root/.helm/plugins
    Creating /root/.helm/starters
    Creating /root/.helm/cache/archive
    Creating /root/.helm/repository/repositories.yaml
    Adding stable repo with URL:
https://aliacs-app-catalog.oss-cn-hangzhou.aliyuncs.com/charts/
    Adding local repo with URL: http://127.0.0.1:8879/charts
    $HELM_HOME has been configured at /root/.helm.
    Not installing Tiller due to 'client-only' flag having been set
    Happy Helming!
```

增加一个国内的阿里云的软件仓库，命令如下：

```
    [root@localhost ~]# helm repo add incubator
https://aliacs-app-catalog.oss-cn-hangzhou.aliyuncs.com/charts-incubator/
    "incubator" has been added to your repositories
    [root@localhost ~]# helm repo update
    Hang tight while we grab the latest from your chart repositories...
    ...Skip local chart repository
    ...Successfully got an update from the "stable" chart repository
    ...Successfully got an update from the "incubator" chart repository
    Update Complete. ⎈ Happy Helming!⎈
```

创建 Helm 服务端，通过-i 选项指定国内 Docker 镜像服务器：

```
    [root@localhost ~]# helm init --service-account tiller --upgrade -i
registry.cn-hangzhou.aliyuncs.com/google_containers/tiller:v3.4.2
--stable-repo-url https://kubernetes.oss-cn-hangzhou.aliyuncs.com/charts
    $HELM_HOME has been configured at /root/.helm.

    Tiller (the Helm server-side component) has been installed into your Kubernetes
Cluster.

    Please note: by default, Tiller is deployed with an insecure 'allow
unauthenticated users' policy.
```

```
    To prevent this, run `helm init` with the --tiller-tls-verify flag.
    For more information on securing your installation see:
https://docs.helm.sh/using_helm/#securing-your-helm-installation
    Happy Helming!
```

启用 TLS 认证，命令如下：

```
    [root@localhost ~]# helm init --service-account tiller --upgrade -i
registry.cn-hangzhou.aliyuncs.com/google_containers/tiller:v3.4.2
--tiller-tls-cert /etc/kubernetes/ssl/tiller001.pem --tiller-tls-key
/etc/kubernetes/ssl/tiller001-key.pem --tls-ca-cert /etc/kubernetes/ssl/ca.pem
--tiller-namespace kube-system --stable-repo-url
https://kubernetes.oss-cn-hangzhou.aliyuncs.com/charts
    $HELM_HOME has been configured at /root/.helm.

    Tiller (the Helm server-side component) has been upgraded to the current
version.
    Happy Helming!
```

由于 Kubernetes 以及 Docker 的很多资源都没有办法通过官方网站直接获得，因此在上面的命令中，我们使用-i 选项指定国内的阿里云的镜像服务器。

实际上 Tiller 是部署在 Kubernetes 集群中的 kube-system 命名空间下的 Deployment，它会去通过调用 kube-api，在 Kubernetes 集群里创建和删除应用。

而从 Kubernetes 1.6 版本开始，API Server 启用了 RBAC 授权。目前的 Tiller 部署时默认没有定义授权的 ServiceAccount，这会导致访问 API Server 时被拒绝。所以我们需要明确为 Tiller 部署添加授权，命令如下：

```
    [root@localhost ~]# kubectl create serviceaccount --namespace kube-system
tiller
    serviceaccount "tiller" created
```

使用以下命令为 Tiller 设置 ServiceAccount：

```
    [root@localhost ~]# kubectl patch deploy --namespace kube-system
tiller-deploy -p '{"spec":{"template":{"spec":{"serviceAccount":"tiller"}}}}'
    "tiller-deploy" patched
```

由于 Tiller 是以 Pod 的形式运行的，因此用户可以使用 kubectl 命令查看其状态，如下所示：

```
    [root@localhost ~]# kubectl get po --namespace=kube-system
    NAME                              READY   STATUS     RESTARTS    AGE
    …
    tiller-deploy-7cb87ddf7d-zxhdt    1/1     Running    4           2d14h
```

由结果可以得知，Tiller 已经处于运行状态，表示已经安装成功。

此时，用户可以通过以下命令查看 Helm 的客户端和服务器的版本：

```
[root@localhost ~]# helm version
    Client: &version.Version{SemVer:"v3.4.2",
GitCommit:"618447cbf203d147601b4b9bd7f8c37a5d39fbb4", GitTreeState:"clean"}
    Server: &version.Version{SemVer:"v3.4.2",
GitCommit:"618447cbf203d147601b4b9bd7f8c37a5d39fbb4", GitTreeState:"clean"}
```

7.3 Chart 文件结构

Helm使用称为Chart的软件包格式。Chart是描述一组相关的Kubernetes资源的文件集合。通过 Chart，管理员可以很方便地在 Kubernetes 中部署应用。本节将介绍 Chart 的结构。

从本质上讲，Chart 软件包实际上就是一个 tar 归档文件。为了研究 Chart 包的具体结构，用户可以从软件仓库上面下载一个已经打包好的 Chart。Helm 提供一个 helm fetch 命令，将 Chart 包从远程软件仓库上下载到本地。例如，下面的命令将 Wordpress 下载到本地计算机中：

```
[root@localhost ~]# helm fetch stable/wordpress
```

下载完成后，查看下载到的文件信息，如下所示：

```
[root@localhost ~]# ll
total 26728
…
-rw-r--r--     1   root    root    14481   Apr  8 08:40 wordpress-0.8.8.tgz
…
```

从上面的输出可知，通过 helm fetch 命令下载到的 Chart 包，其名称为 wordpress-0.8.8.tgz，其中，.tgz 表示该文件是经过 tar 和 gzip 压缩后的文件。

用户可以通过 tar 命令将其解压缩出来，如下所示：

```
[root@localhost ~]# tar -zxvf wordpress-0.8.8.tgz
```

解压缩之后得到的目录名为 wordpress，其结构如下：

```
[root@localhost ~]# ll wordpress
total 36
drwxr-xr-x  3   root   root   21      Apr  8  08:37    charts
-rwxr-xr-x  1   root   root   443     Jan  1  1970     Chart.yaml
-rwxr-xr-x  1   root   root   13062   Jan  1  1970     README.md
-rwxr-xr-x  1   root   root   233     Jan  1  1970     requirements.lock
-rwxr-xr-x  1   root   root   173     Jan  1  1970     requirements.yaml
drwxr-xr-x  3   root   root   206     Apr  8  08:45    templates
-rwxr-xr-x  1   root   root   6768    Jan  1  1970     values.yaml
```

从上面的输出结果可知，一个 Chart 包的内容主要包括 Chart.yaml 和 requirements.yaml 等配置文件，以及 charts 和 templates 等目录。

7.4 使用 Helm

Helm 的使用包括软件仓库管理、查找 Chart、安装 Chart、查看已安装 Chart 以及删除 Chart 等操作，下面分别介绍这些操作的使用方法。

7.4.1 软件仓库的管理

默认情况下，Helm 已经提供了两个软件仓库，一个名为 stable，其网址如下：

https://kubernetes-charts.storage.googleapis.com

另外一个名为 local，其网址如下：

http://127.0.0.1:8879/charts

第一个网址无法访问，管理员通常需要更换默认的软件仓库的网址。

删除软件仓库需要使用 helm repo remove 命令，例如以下命令删除名为 stable 的软件仓库：

```
[root@localhost ~]# helm repo remove stable
```

添加软件仓库可以使用 helm repo add 命令，该命令接受 2 个参数，第 1 个为软件仓库的名称，第 2 个为软件仓库的地址。例如，下面的命令添加国内的阿里云镜像为默认的 stable 软件仓库：

```
[root@localhost ~]# helm repo add stable
https://kubernetes.oss-cn-hangzhou.aliyuncs.com/charts
```

修改完软件仓库的配置之后，需要更新软件仓库的内容，如下所示：

```
[root@localhost ~]# helm repo update
Hang tight while we grab the latest from your chart repositories...
...Skip local chart repository
...Successfully got an update from the "incubator" chart repository
...Successfully got an update from the "stable" chart repository
Update Complete. ⎈ Happy Helming!⎈
```

7.4.2 查找 Chart

搜索远程软件仓库中的 Chart 需要使用 search 命令，如果没有提供命令参数，则会列出所有的 Chart 包，如下所示：

```
[root@localhost ~]# helm search
NAME                  CHART VERSION   APP VERSION     DESCRIPTION
…
incubator/mysql-broker      0.1.0                    A Helm chart for
Kubernetes
```

```
    incubator/mysqlha            0.3.0    5.7.13         MySQL cluster with a
single…
    stable/aerospike         0.1.7    v3.14.1.2      A Helm chart for Aerospike…
    stable/anchore-engine        0.1.3    0.1.6          Anchore container
analysis...
    stable/artifactory           7.0.3    5.8.4          Universal Repository...
    stable/artifactory-ha        0.1.0    5.8.4          Universal Repository...
    stable/bitcoind              0.1.0    0.15.1         Bitcoin is an innovative
payment network and a new kind o...
    …
```

helm search 命令也支持关键字查询，例如，下面的命令查找名称中包含 mysql 的 Chart 包：

```
    [root@localhost ~]# helm search mysql
    NAME                       CHART VERSION   APP VERSION         DESCRIPTION
    incubator/mysql-broker       0.1.0                              A Helm chart for
Kubernetes
    incubator/mysqlha            0.3.0             5.7.13           MySQL cluster
with a single...
    stable/mysql                 0.3.5                              Fast, reliable,
scalable, and..
    stable/percona               0.3.0                              free, fully
compatible,...
    stable/percona-xtradb-cluster  0.0.2           5.7.19           free, fully
compatible,...
    stable/gcloud-sqlproxy       0.2.3                              Google Cloud SQL
Proxy
    stable/mariadb               2.1.6             7.1.31           Fast, reliable,
scalable,...
```

在上面的输出列表中，NAME 为 Chart 的路径名，斜线前面为软件仓库的名称，后面为 Chart 的包名。CHART VERSION 为 Chart 包的版本号，APP VERSION 为 Chart 包的软件版本。例如，incubator/mysqlha 的 Chart 包版本为 0.3.0，该 Chart 包中的 MySQL 的版本号为 5.7.13。

对于某个具体的 Chart 包，用户可以通过 helm inspect 命令查看其详细信息，如下所示：

```
    [root@localhost ~]# helm inspect stable/mysql
    description: Fast, reliable, scalable, and easy to use open-source relational
database
      system.
    engine: gotpl
    home: https://www.mysql.com/
    icon: https://www.mysql.com/common/logos/logo-mysql-170x115.png
    keywords:
    - mysql
    - database
```

```
- sql
maintainers:
- email: viglesias@google.com
  name: Vic Iglesias
name: mysql
sources:
- https://github.com/kubernetes/charts
- https://github.com/docker-library/mysql
version: 0.3.5

---
## mysql image version
## ref: https://hub.docker.com/r/library/mysql/tags/
##
image: "mysql"
imageTag: "5.7.14"

## Specify password for root user
##
## Default: random 10 character string
# mysqlRootPassword: testing

## Create a database user
##
# mysqlUser:
# mysqlPassword:

## Allow unauthenticated access, uncomment to enable
##
# mysqlAllowEmptyPassword: true

## Create a database
##
# mysqlDatabase:
…
```

7.4.3 安装 Chart 包

Chart 包的安装需要使用 helm install 命令，该命令的基本语法如下：

```
helm install [chart] [flags]
```

其中，chart 为要安装的 Chart 包的名称，flags 为一系列的选项。常用的选项有：

- --name：指定 Chart 包的名称。
- --set：该选项为一组"键-值对"，用来覆盖 Chart 包的配置文件默认的配置选项。
- -f：通过 YAML 配置文件覆盖默认选项。

例如，下面的命令安装 stable 仓储中的 mysql：

```
    [root@localhost ~]# helm install --name mysql --set
persistence.storageClass=managed-nfs-storage stable/mysql
    NAME:   mysql
    LAST DEPLOYED: Sun Apr  7 23:49:44 2019
    NAMESPACE: default
    STATUS: DEPLOYED

    RESOURCES:
    ==> v1/PersistentVolumeClaim
    NAME         STATUS  VOLUME  CAPACITY    ACCESS MODES    STORAGECLASS AGE
    mysql-mysql     Pending  managed-nfs-storage                          0s

    ==> v1/Pod(related)
    NAME                        READY        STATUS       RESTARTS       AGE
    mysql-mysql-549d644d4-lmvzl     0/1          Pending      0          0s

    ==> v1/Secret
    NAME               TYPE     DATA     AGE
    mysql-mysql            Opaque   2        0s

    ==> v1/Service
    NAME         TYPE         CLUSTER-IP       EXTERNAL-IP     PORT(S)      AGE
    mysql-mysql     ClusterIP   10.97.230.202   <none>          3306/TCP     0s

    ==> v1beta1/Deployment
    NAME         READY   UP-TO-DATE       AVAILABLE    AGE
    mysql-mysql     0/1     1               0            0s


    NOTES:
    MySQL can be accessed via port 3306 on the following DNS name from within your
cluster:
    mysql-mysql.default.svc.cluster.local

    To get your root password run:

        MYSQL_ROOT_PASSWORD=$(kubectl get secret --namespace default mysql-mysql
-o jsonpath="{.data.mysql-root-password}" | base64 --decode; echo)
```

```
    To connect to your database:

    1. Run an Ubuntu pod that you can use as a client:

        kubectl run -i --tty ubuntu --image=ubuntu:16.04 --restart=Never -- bash
-il

    2. Install the mysql client:

        $ apt-get update && apt-get install mysql-client -y

    3. Connect using the mysql cli, then provide your password:
        $ mysql -h mysql-mysql -p

    To connect to your database directly from outside the K8s cluster:
        MYSQL_HOST=127.0.0.1
        MYSQL_PORT=3306

        # Execute the following commands to route the connection:
        export POD_NAME=$(kubectl get pods --namespace default -l
"app=mysql-mysql" -o jsonpath="{.items[0].metadata.name}")
        kubectl port-forward $POD_NAME 3306:3306

        mysql -h ${MYSQL_HOST} -P${MYSQL_PORT} -u root -p${MYSQL_ROOT_PASSWORD}
```

在上面的命令中，通过--set 选项将新建的 Pod 的 persistence.storageClass 指定为前面创建的 managed-nfs-storage，这样的话，就可以动态为 Pod 提供存储资源了。

接下来，我们验证一下 Kubernetes 的各种资源是否已经自动创建。

首先是检查 Service 是否被创建，如下所示：

```
    [root@localhost ~]# kubectl get svc -o wide
    NAME     TYPE         CLUSTER-IP  EXTERNAL-IP PORT(S) AGE     SELECTOR
    kubernetes       ClusterIP   10.96.0.1        <none>       443/TCP      5d15h
<none>
    mysql-mysql ClusterIP   10.97.230.202   <none>       3306/TCP46m
app=mysql-mysql
```

从上面的输出结果可知，当前集群中已经自动创建了一个名为 mysql-mysql 的 Service，并且其 SELECTOR 为 mysql-mysql。

然后再查看一下 Deployment 的情况，如下所示：

```
    [root@localhost ~]# kubectl get deployments -o wide
    NAME         READY   UP-TO-DATE  AVAILABLE   AGE      CONTAINERS        IMAGES
    SELECTOR
```

```
mysql-mysql     1/1     1         1         49m     mysql-mysql
mysql:5.7.14    app=mysql-mysql
```

从上面的输出可知，一个名为 mysql-mysql 的 Deployment 已经被自动创建，并且其 SELECTOR 定义为 mysql-mysql，这个标签选择器的名称正与前面的 Service 中选择器的名称一致。

然后再查看一下 Pod 的创建情况，如下所示：

```
[root@localhost ~]# kubectl get pod
NAME                                      READY     STATUS     ESTARTS    AGE
mysql-mysql-549d644d4-lmvzl                     1/1     Running     0          35m
nfs-client-provisioner-77d69555bd-hw9md     1/1 Running     2          4d16h
```

在上面的输出结果中，名为mysql-mysql-549d644d4-lmvzl的Pod就是刚才自动创建的Pod。

最后检查一下持久化存储卷请求是否被自动创建，如下所示：

```
[root@localhost ~]# kubectl get pvc
NAME               STATUS                               VOLUME    CAPACITY
ACCESS MODES    STORAGECLASS              AGE
mysql-mysql Bound    pvc-c375de9c-594c-11e9-9c6f-000c29ecaa4e    8Gi
RWO              managed-nfs-storage      58m
```

通过上面的介绍可以看到，通过 Helm，用户可以非常方便地部署一个 MySQL 的应用，并且不需要编辑和管理众多的 YAML 配置。

7.4.4　查看已安装 Chart

Helm 通过 helm ls 命令查看当前系统中已经安装的 Chart 列表，也就是 Release 的列表。如果没有使用选项，该命令默认列出已经成功部署或者失败的 Release，如下所示：

```
    [root@localhost ~]# helm ls
    NAME    REVISION    UPDATED                       STATUS       CHART APP
VERSION     NAMESPACE
    mysql   1           Sun Apr  7 23:49:44 2019     DEPLOYED     mysql-0.3.5
            default
```

如果想要查看所有的 Release，则可以使用-a 选项。通过该选项，可以列出所有状态的 Release，包括被删除的 Release。

```
    [root@localhost ~]# helm list -a
    NAME         REVISION    UPDATED                       STATUS       CHART APP
VERSION    NAMESPACE
    dusty-moose     1        Sun Apr  7 23:03:39 2019     DELETED      mysql-0.3.5
        default
    mysql           1        Sun Apr  7 23:49:44 2019     DEPLOYED     mysql-0.3.5
        default
    …
```

在上面的输出结果中，可以发现第 1 行的 STATUS 为 DELETED，表示该 Release 已经被删除。

7.4.5 删除 Release

对于不再需要的 Release，管理员可以将其删除，删除命令为 helm delete，语法如下：

```
helm delete [flags] release_name
```

具体的示例命令如下所示：

```
[root@localhost ~]# helm delete ill-rottweiler
release "ill-rottweiler" deleted
```

执行以上命令之后，Release 的状态就变成了 DELETED。但是，该 Release 仍然存在于 Kubernetes 系统中，Release 的名称仍然不能被新的 Release 所使用。如果想要彻底将其删除，可以使用--purge 选项，如下所示：

```
[root@localhost ~]# helm delete ill-rottweiler --purge
release "ill-rottweiler" deleted
```

第 8 章

Kubernetes网络管理

Kubernetes 时代的到来，绕不开网络这个话题，网络的基础就是通信。网络之间如何通信、有哪些接口、可选的方案有几种？这些问题将在本章说明。除此之外，读者还需要了解一些 Linux 中的网络专属词汇，这都是 Kubernetes 网络管理的基础。

本章涉及的知识点有：

- Kubernetes 网络基础：主要介绍 Kubernetes 网络模型、Iptables/Netfilter、路由等。
- Kubernetes 网络方案：介绍 Kubernetes 集群中容器到容器的通信以及 Pod 之间的通信等。
- 网络实例分析：通过一个具体的实例来介绍 Kubernetes 集群中从 Pod 到 Service 的网络实现。
- Flannel：主要介绍 CNI 网络模型概念、CNI 网络规范以及 CNI 插件。

8.1 Kubernetes 网络基础

Kubernetes 集群的主要功能是提供各种网络服务。前面的第 7 章和第 8 章中，读者已经初步接触到了 Kubernetes 的网络设置，本节将详细介绍 Kubernetes 的网络基础知识。

8.1.1 Kubernetes 网络模型

在 Kubernetes 中，IP 地址的分配是以 Pod 为单位进行分配的，即每个 Pod 都有一个独立的 IP 地址。而在同一个 Pod 内部，不管有多少个容器，都共享同一个网络命名空间，即 IP 地址、网络设备以及各项网络配置都是共享的。因此，Kubernetes 的网络模型被称为 IP-per-Pod。总的说来，Kubernetes 的网络模型符合以下规则：

（1）在集群中，每个 Pod 都拥有一个独立的 IP 地址，而且假定所有 Pod 都在一个可以直接连通的、扁平的网络空间中，不管是否运行在同一节点上，都可以通过 Pod 的 IP 来访问。

（2）Kubernetes 中 Pod 的 IP 分配是以 Pod 为单位进行的，同一个 Pod 内所有的容器共享一个网络堆栈，该模型称为 IP-per-Pod 模型。

（3）从端口分配、域名解析、服务发现、负载均衡以及应用配置等角度看，Pod 都可以看作是一台独立的虚拟机或者物理机。

（4）所有的容器都可以直接同别的容器通信，而不用通过 NAT。

在 Kubernetes 的网络模型中，有以下几个概念非常重要。

8.1.2 命名空间

命名空间是 Linux 系统中用来隔离资源的一种方式。如果把 Linux 操作系统比作一个大房子，那命名空间指的就是这个房子中的一个个小房间，住在每个房间里的人都自以为独享了整个房子的资源，但其实大家仅仅只是在共享的基础之上互相隔离。共享指的是共享全局的资源，而隔离指的是局部上彼此保持隔离，因而命名空间的本质就是指一种在空间上隔离的概念。当下流行的许多容器虚拟化技术，例如 Docker，就是基于 Linux 命名空间的概念而来的。

网络命名空间用来隔离各种网络资源，例如 IP 地址、路由、网络接口等。后台进程可以运行在不同命名空间内的相同端口上。每个网络命名空间都有自己的路由表，它通过自己的 iptables 配置提供 NAT 和过滤的功能。Linux 网络命名空间还提供了在网络命名空间内运行进程的功能。

8.1.3 veth 网络接口

veth 就是虚拟以太网设备，它都是成对出现的，一端连着网络协议栈，一端彼此相互连接着，如图 8-1 所示。

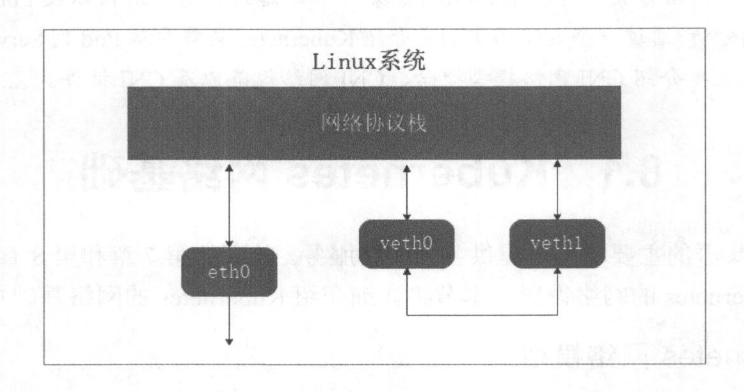

图 8-1 veth 网络接口

正因为有这个特性，veth 常常充当着一个桥梁，连接着各种虚拟网络设备。其中，最常见的情况就是连接两个命名空间（或者容器）等。

在 Linux 系统中，管理员可以通过 ip 命令来查看当前系统中的 veth，如下所示：

```
    [root@localhost ~]# ip link show
    1: lo: <LOOPBACK,UP,LOWER_UP> mtu 65536 qdisc noqueue state UNKNOWN mode
DEFAULT group default qlen 1000
        link/loopback 00:00:00:00:00:00 brd 00:00:00:00:00:00
    2: ens33: <BROADCAST,MULTICAST,UP,LOWER_UP> mtu 1500 qdisc pfifo_fast state
UP mode DEFAULT group default qlen 1000
        link/ether 00:0c:29:ec:aa:4e brd ff:ff:ff:ff:ff:ff
```

```
    3: docker0: <NO-CARRIER,BROADCAST,MULTICAST,UP> mtu 1500 qdisc noqueue state
DOWN mode DEFAULT group default
        link/ether 02:42:fc:eb:a7:0d brd ff:ff:ff:ff:ff:ff
    4: flannel.1: <BROADCAST,MULTICAST,UP,LOWER_UP> mtu 1450 qdisc noqueue state
UNKNOWN mode DEFAULT group default
        link/ether d2:24:3b:4c:66:f6 brd ff:ff:ff:ff:ff:ff
    5: cni0: <BROADCAST,MULTICAST,UP,LOWER_UP> mtu 1450 qdisc noqueue state UP
mode DEFAULT group default qlen 1000
        link/ether 5a:c6:c4:a4:da:4d brd ff:ff:ff:ff:ff:ff
    6: veth45f4a23e@if3: <BROADCAST,MULTICAST,UP,LOWER_UP> mtu 1450 qdisc noqueue
master cni0 state UP mode DEFAULT group default
        link/ether 1e:dd:b2:61:d6:91 brd ff:ff:ff:ff:ff:ff link-netnsid 0
    8: veth0175a568@if3: <BROADCAST,MULTICAST,UP,LOWER_UP> mtu 1450 qdisc noqueue
master cni0 state UP mode DEFAULT group default
        link/ether 9a:c7:42:a0:2f:14 brd ff:ff:ff:ff:ff:ff link-netnsid 2
    9: vethe77105f4@if3: <BROADCAST,MULTICAST,UP,LOWER_UP> mtu 1450 qdisc noqueue
master cni0 state UP mode DEFAULT group default
        link/ether 0e:dd:5c:0f:59:59 brd ff:ff:ff:ff:ff:ff link-netnsid 1
```

在上面的输出中，编号为 6、8 和 9 的网络设备即 veth。

8.1.4　netfilter/iptables

netfilter 负责在内核中执行各种数据包过滤规则，运行在内核模式中。iptables 是在用户模式下运行的进程，负责协助维护一系列数据包过滤规则表，通过二者的配合来实现整个 Linux 网络协议栈中灵活的数据包处理机制。

8.1.5　网桥

网桥是一个二层网络设备，通过网桥可以将 Linux 支持的不同的端口连接起来，并实现类似交换机那样的多对多的通信。

8.1.6　路由

Linux 系统包含一个完整的路由功能，当 IP 层在处理数据发送或转发的时候，会使用路由表来决定发往哪里。在 CentOS 中，用户可以通过 ip route 命令查看当前相同的路由表，如下所示：

```
    [root@localhost ~]# ip route show
    default via 192.168.21.2 dev ens33 proto dhcp metric 100
    172.17.0.0/16 dev docker0 proto kernel scope link src 172.17.0.1
    192.168.21.0/24 dev ens33 proto kernel scope link src 192.168.21.135 metric
100
```

其中，第 2 条规则的意思是所有 docker0 接口上收到的、发往 172.17.0.0/16 网段的数据包都由本机处理。

8.2 Kubernetes 网络实现

在实际应用场景中，Kubernetes 集群的网络拓扑比较复杂，涉及 Pod 内部的容器之间、Pod 之间、节点之间、Service 与 Pod 之间以及集群与外部网络之间的通信。本节将详细介绍这些通信的实现方式。

8.2.1 Docker 与 Kubernetes 网络比较

图 8-2 展示了 Kubernetes 集群中的三种 IP 地址和三种网络的概念。为了便于理解，可以将节点的 IP 地址比作 TCP/IP 网络中的第二层地址，即 MAC 地址，通过它去寻找物理节点。而这个地址通常对 Kubernetes 里的 Pod 来说是透明的，不用知道其他 Pod 的节点 IP 地址，通过 Pod 的 IP 地址就能访问到。所以 Pod IP 地址可以看作是网络结构中的三层 IP 地址。而 ClusterIP 更像是一个域名，不用知道背后到底有哪些 Pod，它们又分布在哪里。

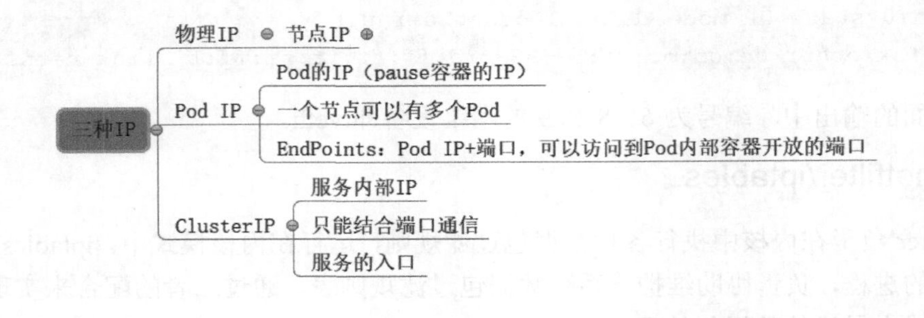

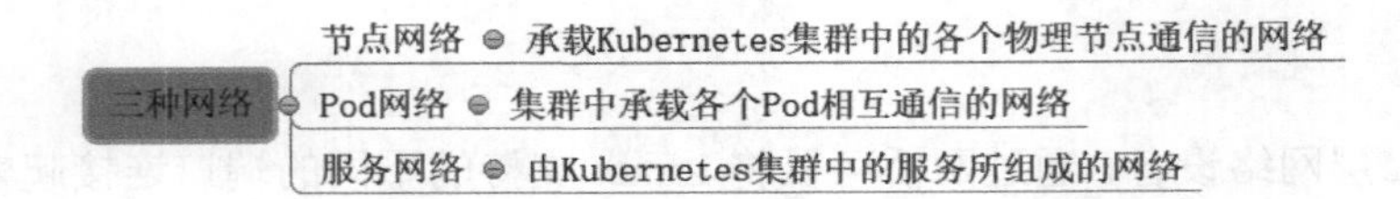

图 8-2　Docker 与 Kubernetes 网络比较

1. Docker 网络实现

用过 Docker 基本都知道，启动 docker engine 后，主机的网络设备里会有一个 docker0 的网关，而容器默认情况下会被分配在一个以 docker0 为网关的虚拟子网中。

```
[root@localhost ~]# ip address show
1: lo: <LOOPBACK,UP,LOWER_UP> mtu 65536 qdisc noqueue state UNKNOWN group default qlen 1000
    link/loopback 00:00:00:00:00:00 brd 00:00:00:00:00:00
    inet 127.0.0.1/8 scope host lo
      valid_lft forever preferred_lft forever
    inet6 ::1/128 scope host
      valid_lft forever preferred_lft forever
```

```
    2: ens33: <BROADCAST,MULTICAST,UP,LOWER_UP> mtu 1500 qdisc pfifo_fast state
UP group default qlen 1000
        link/ether 00:0c:29:b3:76:16 brd ff:ff:ff:ff:ff:ff
        inet 192.168.21.137/24 brd 192.168.21.255 scope global noprefixroute ens33
          valid_lft forever preferred_lft forever
        inet6 fe80::ce7b:a6d7:32d0:e630/64 scope link noprefixroute
          valid_lft forever preferred_lft forever
    3: docker0: <BROADCAST,MULTICAST,UP,LOWER_UP> mtu 1500 qdisc noqueue state
UP group default
        link/ether 02:42:da:e7:0b:a3 brd ff:ff:ff:ff:ff:ff
        inet 172.17.2.1/24 scope global docker0
          valid_lft forever preferred_lft forever
        inet6 fe80::42:daff:fee7:ba3/64 scope link
          valid_lft forever preferred_lft forever
```

实际上，docker0 是一个虚拟网桥，工作在第二层网络。也可以为它配置 IP，工作在三层网络。通过 docker0，将各个容器连接起来。

2．Kubernetes 网络实现

Docker 默认的网络是为同一台宿主机的 Docker 容器通信设计的，Kubernetes 的 Pod 需要跨主机与其他 Pod 通信，所以需要设计一套让不同节点的 Pod 实现透明通信，即不通过 NAT 的机制。

docker0 的默认 IP 地址是 172.17.2.1，Docker 启动的容器也默认被分配在 172.17.2.1/24 的网段里。跨主机的 Pod 通信要保证 Pod 的 IP 地址不能相同，所以还需要设计一套为 Pod 统一分配 IP 地址的机制。

下面的命令显示了同一个 Pod 里面的不同容器的信息。其中，pause 容器的信息如下：

```
    [root@localhost ~]# docker inspect 3b393a9399ce
    [
        {
            "Id":
"3b393a9399ce0449009226396a284e3ec34a9f1d2b089f43834691537ef19978",
            "Created": "2019-04-14T09:16:49.31267411Z",
            "Path": "/usr/bin/pod",
            "NetworkSettings": {
                "Bridge": "",
                "SandboxID":
"c442d814bbd33ab591bbc5f5723854e83f82a6f0ce217f19a3aeb47f02969e1a",
                "HairpinMode": false,
                "LinkLocalIPv6Address": "",
                "LinkLocalIPv6PrefixLen": 0,
                "Ports": {},
                "SandboxKey": "/var/run/docker/netns/c442d814bbd3",
```

```
                "SecondaryIPAddresses": null,
                "SecondaryIPv6Addresses": null,
                "EndpointID":
"e8b14fbc821772e575a60d94bfe8de5ea0970db9c9c61633ffa6c97bdbd9ee96",
                "Gateway": "172.17.2.1",
                "GlobalIPv6Address": "",
                "GlobalIPv6PrefixLen": 0,
                "IPAddress": "172.17.2.2",
                "IPPrefixLen": 24,
                "IPv6Gateway": "",
                "MacAddress": "02:42:ac:11:02:02",
                "Networks": {
                    "bridge": {
                        "IPAMConfig": null,
                        "Links": null,
                        "Aliases": null,
                        "NetworkID":
"d73b68325814201b064bcbd5903bcf01c34f778620172a1ad229792292dffd4f",
                        "EndpointID":
"e8b14fbc821772e575a60d94bfe8de5ea0970db9c9c61633ffa6c97bdbd9ee96",
                        "Gateway": "172.17.2.1",
                        "IPAddress": "172.17.2.2",
                        "IPPrefixLen": 24,
                        "IPv6Gateway": "",
                        "GlobalIPv6Address": "",
                        "GlobalIPv6PrefixLen": 0,
                        "MacAddress": "02:42:ac:11:02:02"
                    }
                }
            }
```

而另外一个普通容器的信息如下：

```
    [root@localhost ~]# docker inspect c34c09ad53f3
    [
        {
            "Id":
"c34c09ad53f3b636c0c65567041355713c03ef207fb955482a0c8bd3c3676878",
            "Created": "2019-04-14T09:17:01.972754623Z",
            "Path": "catalina.sh",
            "Args": [
                "run"
            ],
            "State": {
```

```
            "Status": "running",
            "Running": true,
            "Paused": false,
            "Restarting": false,
            "OOMKilled": false,
            "Dead": false,
            "Pid": 17098,
            "ExitCode": 0,
            "Error": "",
            "StartedAt": "2019-04-14T09:17:02.282104422Z",
            "FinishedAt": "0001-01-01T00:00:00Z"
        },
        "Image":
"sha256:5a069ba3df4d4221755d76d905ce8a0d2eedf3edbd87dca05a6259114c7b93d4",
        "ResolvConfPath": "/var/lib/docker/containers/3
b393a9399ce0449009226396a284e3ec34a9f1d2b089f43834691537ef19978/resolv.conf",
        "HostnamePath": "/var/lib/docker/containers/
3b393a9399ce0449009226396a284e3ec34a9f1d2b089f43834691537ef19978/hostname",
        "HostsPath": "/var/lib/kubelet/pods/f951d250-5e2c-11e9-94d8-
000c29ce2559/etc-hosts",
        "LogPath": "",
        "Name":
"/k8s_tomcat.cdbc0245_frontend_default_f951d250-5e2c-11e9-94d8-000c29ce2559_4
bca6f7c",
        "RestartCount": 0,
        "Driver": "overlay2",
        "MountLabel": "",
        "ProcessLabel": "",
        "AppArmorProfile": "",
        "ExecIDs": null,
        "HostConfig": {
            "Binds": [
                "/var/lib/kubelet/pods/f951d250-5e2c-11e9-94d8-000c29ce2559/
etc-hosts:/etc/hosts",
                "/var/lib/kubelet/pods/f951d250-5e2c-11e9-94d8-000c29ce2559/
containers/tomcat/4bca6f7c:/dev/termination-log"
            ],
            "ContainerIDFile": "",
            "LogConfig": {
                "Type": "journald",
                "Config": {}
            },
```

```
            "NetworkMode":
"container:3b393a9399ce0449009226396a284e3ec34a9f1d2b089f43834691537ef19978",
            "PortBindings": null,
```

从上面的命令可以看出，在这个 Pod 中，普通的容器通过 NetworkMode 字段与 pause 容器共享了网络。在这种情况下，同一个 Pod 里面的容器相互之间的访问，只需要通过 localhost 加端口的形式就可以了。

更进一步地思考一下，pause 容器的 IP 地址又是从哪里获取到的？如果还是以 docker0 为网关的内网 IP，就会出现问题了。

docker0 的默认 IP 地址是 172.17.2.1，Docker 启动的容器也默认被分配在 172.17.2.0/24 网段里。跨主机的 Pod 通信要保证 Pod 的 IP 地址不能相同，所以还需要设计一套为 Pod 统一分配 IP 的机制。

以上就是 Kubernetes 在 Pod 网络这一层需要解决的问题。目前 Kubernetes 提供了许多网络插件来解决这个问题，比较常见的有 Flannel、CNI 以及 DANM 等。

8.2.2 容器之间的通信

前面已经介绍过，Pod 是 Kubernetes 集群中最小的调度单元。而 Pod 实际上是容器的集合，在 Pod 中可以包含一个或者多个容器。Pod 包含的容器都运行在一个节点上，这些容器拥有相同的网络空间，容器之间能够相互通信。Pod 网络本质上还是容器网络，所以 Pod 的 IP 地址就是 Pod 中第一个容器的 IP 地址。

Docker 云的网络模型为一个扁平化网络，Pod 作为一个网络单元同 Kubernetes 节点的网络处于同一层级。

同一个 Pod 之间的不同容器因为共享同一个网络命名空间，所以可以直接通过 localhost 进行通信。

假设在当前集群中存在着一个 Pod，其配置文件如下：

```
apiVersion: v1
kind: Pod
metadata:
  name: two-containers
spec:
  restartPolicy: Never
  volumes:
  - name: shared-data
    emptyDir: {}

  containers:

  - name: nginx-container
    image: nginx
    volumeMounts:
```

```
      - name: shared-data
        mountPath: /usr/share/nginx/html
    - name: mysql-container
      image: mysql
      env:
      - name: MYSQL_ROOT_PASSWORD
        value: a123456
    - name: centos-container
      image: centos
      command: [ "/bin/bash", "-c", "--" ]
      args: [ "while true; do sleep 30; done;" ]
```

从上面的代码可知，该 Pod 包含 3 个容器，分别为 nginx-container、mysql-container 和 centos-container。

创建以上 Pod 之后，执行以下命令进入到 centos-container 容器中：

```
[root@localhost ~]# kubectl exec -it two-containers -c centos-container --
/bin/sh
```

然后执行以下命令来测试是否可以访问到 nginx-container：

```
sh-4.2# curl http://localhost:80
<html>
<head><title>403 Forbidden</title></head>
<body>
<center><h1>403 Forbidden</h1></center>
<hr><center>nginx/1.15.10</center>
</body>
</html>
```

前面已经介绍过，curl 命令是一个功能强大的 HTTP 客户端工具。在上面的例子中，通过该命令访问 nginx-container 的 80 端口。可以发现，上面的输出代码正是 Nginx 的输出结果。从上面的例子可知，用户可以在同一个 Pod 的容器中，通过 localhost 和端口来直接访问另外一个容器。

接下来，我们在同一个容器中访问 mysql-container。这次使用 telnet 命令，如下所示：

```
sh-4.2# telnet localhost 3306
Trying ::1...
Connected to localhost.
Escape character is '^]'.
J
8.0.15
     AI8@IU▒:)A~y3l4caching_sha2_password
```

从上面的输出结果可知，centos-container 同样也可以通过 localhost 加 3306 端口直接访问 mysql-container 中的服务。

通过上面的例子，可以充分说明，同一个 Pod 中的容器是共享一个 IP 地址，用户可以通过端口来区分不同的服务。

8.2.3 Pod 之间的通信

接下来，再讨论一下 Pod 之间的通信。Pod 之间的网络通信主要有两种情况，一种是同一个节点上的不同 Pod 之间的通信，另外一种情况是不同节点上的 Pod 之间的通信。

在同一个节点中，不同的 Pod 都拥有一个全局 IP 地址，Pod 之间可以直接通过 IP 地址进行通信。

例如，在当前节点中，一共有 4 个 Pod，如下所示：

```
    [root@localhost ~]# kubectl get pods -o wide
    NAME                     READY   STATUS   RESTARTS AGE       IP         NODE
    centos-controller-qj0fc      1/1      Running    0         2m        172.17.0.5
127.0.0.1
    nginx-controller-36wdx       1/1      Running    0         8m        172.17.0.4
127.0.0.1
    nginx-controller-9md9b       1/1      Running    0         8m        172.17.0.3
127.0.0.1
    two-containers               3/3      Running    0         8m        172.17.0.2
127.0.0.1
```

这 4 个 Pod 的 IP 地址分别为 172.17.0.2~172.17.0.5。下面先进入名为 centos-controller -qj0fc 的容器，命令如下：

```
    [root@localhost ~]# kubectl exec -it centos-controller-qj0fc -c centos --
/bin/sh
```

然后通过 ping 命令测试是否可以访问 172.17.0.2 的 Pod，执行结果如下：

```
    sh-4.2# ping 172.17.0.2
    PING 172.17.0.2 (172.17.0.2) 56(84) bytes of data.
    64 bytes from 172.17.0.2: icmp_seq=1 ttl=64 time=0.068 ms
    64 bytes from 172.17.0.2: icmp_seq=2 ttl=64 time=0.064 ms
    64 bytes from 172.17.0.2: icmp_seq=3 ttl=64 time=0.051 ms
    64 bytes from 172.17.0.2: icmp_seq=4 ttl=64 time=0.046 ms
    …
```

以上命令的输出结果表明，这 2 个 Pod 之间是连通的。此时，如果 ping 其他 Pod，也会得到类似的结果。

实际上，在同一个节点中，不同的 Pod 通过一个名为 docker0 的网桥连接起来。例如，用户可以通过以下命令将当前节点的网络接口罗列出来：

```
    [root@localhost ~]# ip address show
    1: lo: <LOOPBACK,UP,LOWER_UP> mtu 65536 qdisc noqueue state UNKNOWN group
default qlen 1000
```

```
        link/loopback 00:00:00:00:00:00 brd 00:00:00:00:00:00
        inet 127.0.0.1/8 scope host lo
           valid_lft forever preferred_lft forever
        inet6 ::1/128 scope host
           valid_lft forever preferred_lft forever
    2: ens33: <BROADCAST,MULTICAST,UP,LOWER_UP> mtu 1500 qdisc pfifo_fast state
UP group default qlen 1000
        link/ether 00:0c:29:ce:25:59 brd ff:ff:ff:ff:ff:ff
        inet 192.168.21.135/24 brd 192.168.21.255 scope global noprefixroute
dynamic ens33
           valid_lft 1786sec preferred_lft 1786sec
        inet6 fe80::7b0:8324:3573:789/64 scope link noprefixroute
           valid_lft forever preferred_lft forever
    3: docker0: <BROADCAST,MULTICAST,UP,LOWER_UP> mtu 1500 qdisc noqueue state
UP group default
        link/ether 02:42:b3:b0:7c:3d brd ff:ff:ff:ff:ff:ff
        inet 172.17.0.1/16 scope global docker0
           valid_lft forever preferred_lft forever
        inet6 fe80::42:b3ff:feb0:7c3d/64 scope link
           valid_lft forever preferred_lft forever
```

从上面的输出可知，编号为 3 的 docker0 的 IP 地址为 172.17.0.1，这个 IP 地址与 Pod 的 IP 地址位于同一个网段中。

brctl 命令则可以将网桥的信息显示出来，如下所示：

```
[root@localhost ~]# brctl show
bridge name     bridge id               STP enabled     interfaces
docker0         8000.0242b3b07c3d       no              veth0f8b70e
                                                        veth452dff6
                                                        veth4a2f85a
                                                        veth67fa727
```

从上面的输出结果可知，在当前节点中，有 4 个虚拟网络接口被桥接到了 docker0 上面。正是通过这种方式，实现了 Pod 之间的直接通信。

图 8-3 描述了同一个节点中不同的 Pod 之间的网络通信。从图中可以看出，Pod1 和 Pod2 都是通过 veth 虚拟网络接口连接到同一个 docker0 网桥上，这些虚拟网络接口的 IP 地址都是从 docker0 的网段上动态获取的，它们和 docker0 网桥同属于一个网段，因此，Pod1 和 Pod2 以及 docker0 可以直接通信。

对于不同的节点的 Pod 来说，情况就比较复杂了。不同的节点之间，节点的 IP 地址相当于外网 IP 地址，它们之间可以直接相互访问。但是节点内部的 Pod 和 docker0 网桥的 IP 地址则是内网 IP 地址，无法直接跨越节点访问。如果它们之间想要实现通信，就必须通过节点的网络接口进行转发，如图 8-4 所示。

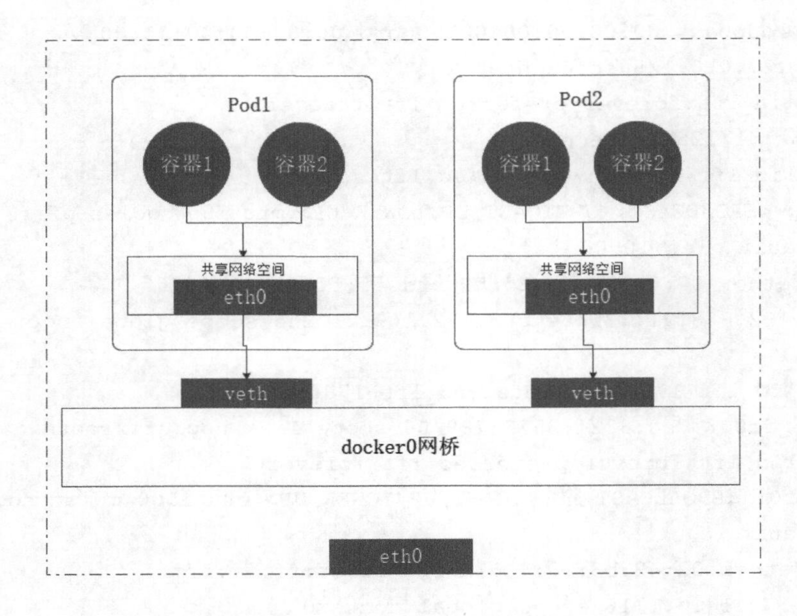

图 8-3　节点内部 Pod 之间的通信

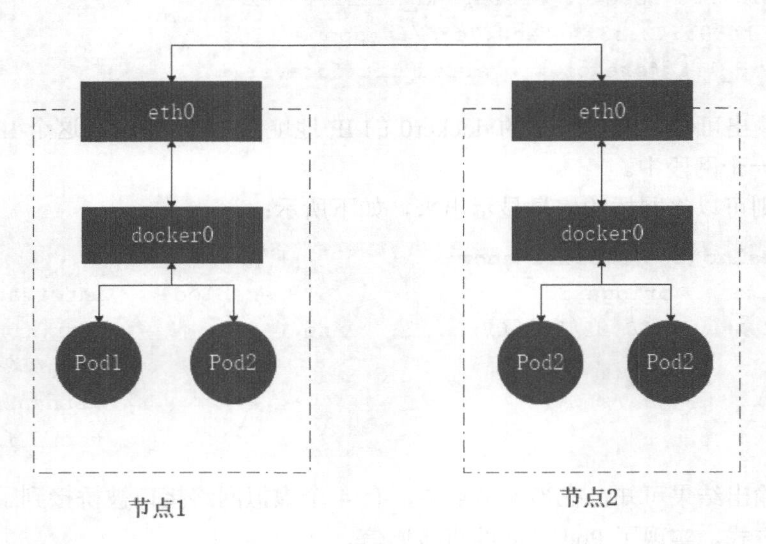

图 8-4　不同节点的 Pod 之间通信

8.2.4　Pod 和服务之间的通信

关于 Pod 和服务之间的关联，实际上在前面的一些章节中已经做了部分介绍。但是，在前面的介绍中，主要内容放在了介绍如何配置和发布服务上面，对于其中的网络原理并没有过多地介绍。在本节中，将从 Kubernetes 的基本网络管理着手，详细介绍 Kubernetes 是如何实现从 Service 到 Pod 中应用系统的访问的。

首先准备 3 个节点，其中一个为 Master 节点，另外 2 个为 Node 节点，其拓扑结构如图 8-5 所示。

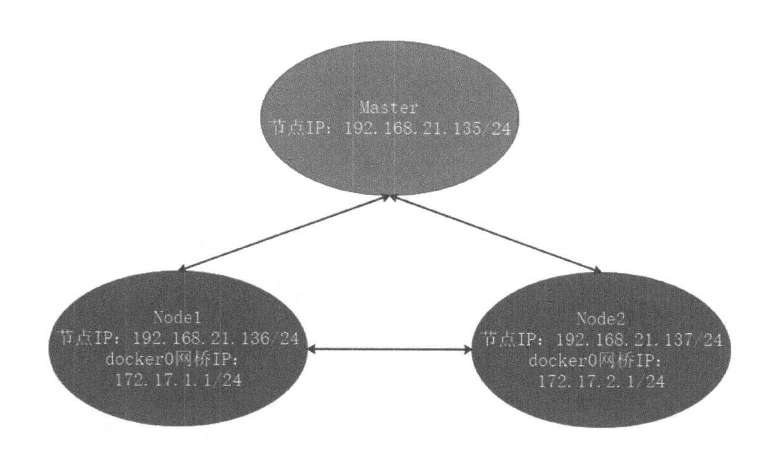

图 8-5　拓扑结构

在图 8-5 中，Master 节点的 IP 地址为 192.168.21.135，Node1 节点的 IP 地址为 192.168.21.136，其中 docker0 网桥的 IP 地址手工设置为 172.17.1.1。Node2 节点的 IP 地址设置为 192.168.21.137，docker0 网桥的 IP 地址设置为 172.17.2.1。通过这样的设置，可以使读者更加深入地理解和掌握 Kubernetes 的网络原理。

首先，在 Node1 和 Node2 这两个节点上修改一下 docker0 网桥的 IP 地址。前面已经介绍过，Docker 默认为 docker0 网络分配了一个 IP 地址 172.17.0.1/16。随后所有的当前节点的容器的 IP 地址都是从 docker0 所在的 172.17.0.0/16 网络中自动分配。当然，在 Docker 中，这个网段仅限于当前节点中访问，是不可以被路由的。所以，尽管在每个节点中，docker0 的 IP 地址以及容器的 IP 地址都属于 172.17.0.0/16 网络，但是用户不必考虑冲突的问题。

而在 Kubernetes 中，所有节点的 docker0 网桥都是可以被路由的，即不同节点之间可以直接相互访问，而不必通过 NAT。为此，我们需要将其修改成不同的网段。

对于 Node1 节点，docker0 网桥的 IP 地址可以直接使用以下命令进行修改：

```
[root@localhost ~]# ip addr add 172.17.1.1/24 dev docker0
```

然后将原来的 IP 地址删除，命令如下：

```
[root@localhost ~]# ip addr delete 172.17.0.1/16 dev docker0
```

使用以下命令查看 IP 地址是否设置成功：

```
[root@localhost ~]# ip a show docker0
3: docker0: <BROADCAST,MULTICAST,UP,LOWER_UP> mtu 1500 qdisc noqueue state
UP group default
    link/ether 02:42:f1:3d:19:89 brd ff:ff:ff:ff:ff:ff
    inet 172.17.1.1/24 scope global docker0
       valid_lft forever preferred_lft forever
    inet6 fe80::42:f1ff:fe3d:1989/64 scope link
       valid_lft forever preferred_lft forever
```

如果输出信息如上所示，则表示已经成功修改。

但是以上命令仅仅是临时生效，当节点被重新启动之后，所有的配置都将丢失。为了能够将配置长久保存下来，用户可以修改 Docker 的配置文件/etc/docker/daemon.json。在配置文件中增加以下代码：

```
{
  "bip": "172.17.1.1/24"
}
```

其中，bip 表示将 docker0 网桥的 IP 地址设置为其后面的值。

用户需要在 Node2 节点上修改/etc/docker/daemon.json 配置文件，增加以下代码：

```
{
   "bip": "172.17.2.1/24"
}
```

通过以上配置，Node1 和 Node2 中容器就会分别赋予 172.17.1.0/24 和 172.17.2.0/24 这两个网络的 IP 地址，并且默认的网关将被分别设置为 docker0 的 IP 地址，即 172.17.1.1 和 172.17.2.1。

修改完成之后，重新启动 Docker 服务。此时，在 Node1 节点上通过 iptables-save 命令查看防火墙规则，会发现在 NAT 表中多出以下规则：

```
-A POSTROUTING -s 172.17.1.0/24 ! -o docker0 -j MASQUERADE
```

以上规则表示源地址为 172.17.1.0/24 的，但是又不是由 docker0 网桥发出的，实际上就是容器发出的数据包，需要进行源地址转换，转换为节点的 IP 地址。

同时，查看 Node1 节点的路由表，也会出现通向 172.17.1.0/24 网络的路由规则，如下所示：

```
[root@localhost ~]# ip route list
default via 192.168.21.2 dev ens33 proto static metric 100
172.17.1.0/24 dev docker0 proto kernel scope link src 172.17.1.1
192.168.21.0/24 dev ens33 proto kernel scope link src 192.168.21.136 metric
100
```

以上信息表明，Kubernetes 已经准备好了 Pod 网络的路由。

注　意

在 Node2 节点上面，用户也可以得到类似的信息。

接下来，继续在集群中部署 Tomcat 应用。在 Master 节点中创建 Tomcat 应用的 YAML 配置文件，如下所示：

```
apiVersion: v1
kind: Pod
metadata:
```

```
   name: tomcat
   labels:
     name:  tomcat
 spec:
   containers:
   - name: tomcat
     image: docker.io/tomcat
     ports:
     - containerPort: 8080
```

将以上代码保存为 tomcat-pod.yaml，

```
[root@localhost ~]# kubectl create -f tomcat-pod.yaml
pod "tomcat" created
```

查看 Pod 的状态，命令如下：

```
[root@localhost ~]# kubectl get pod -o wide
NAME          READY    STATUS   RESTARTS     AGE      IP            NODE
…
tomcat  1/1     Running    0           13s   172.17.1.3    192.168.21.136
```

可以发现，我们刚才创建的 Pod 已经处于运行状态，位于 Node1 节点上，分配给它的 IP 地址为 172.17.1.3。

在 Node1 上查看网络接口，可以发现多出一个 veth 网络接口，如下所示：

```
    [root@localhost ~]# ip link show
    1: lo: <LOOPBACK,UP,LOWER_UP> mtu 65536 qdisc noqueue state UNKNOWN mode
DEFAULT group default qlen 1000
        link/loopback 00:00:00:00:00:00 brd 00:00:00:00:00:00
    2: ens33: <BROADCAST,MULTICAST,UP,LOWER_UP> mtu 1500 qdisc pfifo_fast state
UP mode DEFAULT group default qlen 1000
        link/ether 00:0c:29:4b:2b:04 brd ff:ff:ff:ff:ff:ff
    3: docker0: <BROADCAST,MULTICAST,UP,LOWER_UP> mtu 1500 qdisc noqueue state
UP mode DEFAULT group default
        link/ether 02:42:77:03:ed:43 brd ff:ff:ff:ff:ff:ff
    7: vethbc262f0@if6: <BROADCAST,MULTICAST,UP,LOWER_UP> mtu 1500 qdisc noqueue
master docker0 state UP mode DEFAULT group default
        link/ether 36:b0:c8:a1:f7:91 brd ff:ff:ff:ff:ff:ff link-netnsid 1
```

而该网络接口被桥接到了 docker0 网桥上，如下所示：

```
[root@localhost ~]# brctl show
bridge name    bridge id              STP enabled    interfaces
docker0        8000.02427703ed43      no             vethbc262f0
```

在 Node1 节点上，通过 curl 命令尝试访问 172.17.1.3 的 Pod 的 8080 端口，结果如下：

```
[root@localhost ~]# curl 172.17.1.3:8080



<!DOCTYPE html>
<html lang="en">
    <head>
        <meta charset="UTF-8" />
        <title>Apache Tomcat/8.5.40</title>
        <link href="favicon.ico" rel="icon" type="image/x-icon" />
        <link href="favicon.ico" rel="shortcut icon" type="image/x-icon" />
        <link href="tomcat.css" rel="stylesheet" type="text/css" />
    </head>

    <body>
        <div id="wrapper">
            <div id="navigation" class="curved container">
                <span id="nav-home"><a href="https://tomcat.apache.org/">
Home</a></span>
                <span id="nav-hosts"><a href="/docs/">Documentation</a>
</span>
                <span id="nav-config"><a href="/docs/config/">
Configuration</a></span>
                <span id="nav-examples"><a href="/examples/">
Examples</a></span>
                <span id="nav-wiki"><a href="https://wiki.apache.org/
tomcat/FrontPage">Wiki</a></span>
                <span id="nav-lists"><a href="https://tomcat.apache.org/
lists.html">Mailing Lists</a></span>
                <span id="nav-help"><a href="https://tomcat.apache.org/
findhelp.html">Find Help</a></span>
                <br class="separator" />
            </div>
            <div id="asf-box">
                <h1>Apache Tomcat/8.5.40</h1>
            </div>
            <div id="upper" class="curved container">
                <div id="congrats" class="curved container">
                    <h2>If you're seeing this, you've successfully installed
Tomcat. Congratulations!</h2>
                </div>
...
```

以上的输出结果正是 Tomcat 的默认输出结果，这表明 Tomcat 已可被正常访问。

但是，如果用户在 Master 和 Node2 节点上访问 172.17.1.3:8080，则会出现以下问题：

```
[root@localhost ~]# curl 172.17.1.3:8080
curl: (7) Failed connect to 172.17.1.3:8080; No route to host
```

以上信息表明，在其他的两个节点上，没有发现到 172.17.1.0/24 网络的路由。如果通过 ping 命令分别在 2 个节点上测试到 172.17.1.0/24 网络的网关，即 Node1 上的 docker0 的 IP 地址 172.17.1.1 的连通性，则会输出以下结果：

```
[root@localhost ~]# ping 172.17.1.1
PING 172.17.1.1 (172.17.1.1) 56(84) bytes of data.
From 172.17.0.1 icmp_seq=1 Destination Host Unreachable
From 172.17.0.1 icmp_seq=2 Destination Host Unreachable
From 172.17.0.1 icmp_seq=3 Destination Host Unreachable
…
```

这也表明无法在其余的 2 个节点上访问该网关。

在 Master 节点和 Node2 节点上分别查看路由表信息。其中 Master 节点的路由表如下所示：

```
[root@localhost ~]# ip route list
default via 192.168.21.2 dev ens33 proto dhcp metric 100
172.17.0.0/16 dev docker0 proto kernel scope link src 172.17.0.1
192.168.21.0/24 dev ens33 proto kernel scope link src 192.168.21.135 metric
100
```

Node2 节点的路由表如下：

```
[root@localhost ~]# ip route list
default via 192.168.21.2 dev ens33 proto static metric 100
172.17.2.0/24 dev docker0 proto kernel scope link src 172.17.2.1
192.168.21.0/24 dev ens33 proto kernel scope link src 192.168.21.137 metric
100
```

从上面的输出结果可知，这 2 个节点确实没有访问 172.17.1.0/24 网络的路由规则。

为了能够实现 Pod 的直接访问，用户需要在这 3 个节点上面分别添加到 172.17.1.0/24 和 172.17.2.0/24 这 2 个网络的路由规则。其中，在 Master 节点上，用户需要将这 2 条规则同时添加上去。

```
[root@localhost ~]# ip route add 172.17.1.0/24 via 192.168.21.136
[root@localhost ~]# ip route add 172.17.2.0/24 via 192.168.21.137
```

而在 Node1 和 Node2 节点上，由于本身已经分别有了到 172.17.1.0/24 和 172.17.2.0/24 的路由规则，只需要添加另外一条即可。

到目前为止，用户理论上应该可以在三个节点上都可以访问刚才部署的 Tomcat 应用了。但是实际上还不可以。这是因为除了 Kubernetes 本身的规则之外，默认情况下 Linux 本身的防火墙规则中，INPUT 和 FORWARD 这 2 个链的默认规则都是拒绝的，所以用户需要在 Node1 和 Node2 上分别执行以下命令，将其默认规则设置为 ACCEPT：

```
iptables -P INPUT ACCEPT
iptables -P FORWARD ACCEPT
iptables -F
iptables -L -n
```

最后，用户就会发现，在任意一个节点上都可以直接访问到了 172.17.1.3:8080 这个应用。

下面再接着创建服务，其配置文件如下：

```
apiVersion: v1
kind: Service
metadata:
  name: tomcat
  labels:
    name: tomcat
spec:
  selector:
    name: tomcat
  ports:
  - port: 8080
```

将以上代码保存为 tomcat-service.yaml，然后在 Master 节点上执行以下命令创建该服务：

```
[root@localhost ~]# kubectl create -f tomcat-service.yaml
service "tomcat" created
```

查看刚才创建的服务的状态，如下所示：

```
[root@localhost ~]# kubectl get svc -o wide
NAME         CLUSTER-IP       EXTERNAL-IP PORT(S)        AGE      SELECTOR
kubernetes        10.254.0.1       <none>        443/TCP     4d       <none>
tomcat            10.254.12.13     <none>        8080/TCP    29s      name=tomcat
```

从以上命令的输出可知，Kubernetes 已经为名为 tomcat 的服务分配了一个 ClusterIP，其值为 10.254.12.13，端口为 8080，选择器为 name=tomcat。

ClusterIP 是一个虚拟的 IP 地址，并不在集群或者节点中真实存在。这个 IP 地址是在 kube-apiserver 的配置文件中定义的，如下所示：

```
[root@localhost ~]# cat /etc/kubernetes/apiserver
###
# kubernetes system config
#
# The following values are used to configure the kube-apiserver
#


# The address on the local server to listen to.
KUBE_API_ADDRESS="--insecure-bind-address=0.0.0.0"
```

```
    # The port on the local server to listen on.
    KUBE_API_PORT="--port=8080"

    # Port minions listen on
    # KUBELET_PORT="--kubelet-port=10250"

    # Comma separated list of nodes in the etcd cluster
    KUBE_ETCD_SERVERS="--etcd-servers=http://127.0.0.1:2379"

    # Address range to use for services
    KUBE_SERVICE_ADDRESSES="--service-cluster-ip-range=10.254.0.0/16"

    # default admission control policies
    KUBE_ADMISSION_CONTROL="--admission-control=NamespaceLifecycle,NamespaceE
xists,LimitRanger,ResourceQuota"

    # Add your own!
    KUBE_API_ARGS=""
```

其中--service-cluster-ip-range 选项定义了服务所在的网络。

实际上，ClusterIP 所在的网络可以随意分配，只要不跟物理网络和 docker0 的网络冲突即可。ClusterIP 应用范围仅仅局限于当前节点，不会在物理网络和 docker0 所在的网络上路由。ClusterIP 的作用仅仅是将访问该服务的流量发送到与其绑定的 Endpoints 上。

在任何一个节点上使用 iptables-save 命令查看 iptables 的规则，会发现多出多条与 ClusterIP 有关的规则，如下所示：

```
    01 -A KUBE-SEP-TMJQS2NVDIZNCSFJ -s 172.17.1.3/32 -m comment --comment
"default/tomcat:" -j KUBE-MARK-MASQ
    02 -A KUBE-SEP-TMJQS2NVDIZNCSFJ -p tcp -m comment --comment "default/tomcat:"
-m tcp -j DNAT --to-destination 172.17.1.3:8080
    03 -A KUBE-SEP-WSAW6HE4TRTZJXCF -s 192.168.21.135/32 -m comment --comment
"default/kubernetes:https" -j KUBE-MARK-MASQ
    04 -A KUBE-SEP-WSAW6HE4TRTZJXCF -p tcp -m comment --comment
"default/kubernetes:https" -m recent --set --name KUBE-SEP-WSAW6HE4TRTZJXCF
--mask 255.255.255.255 --rsource -m tcp -j DNAT --to-destination
192.168.21.135:6443
    05 -A KUBE-SERVICES -d 10.254.12.13/32 -p tcp -m comment --comment
"default/tomcat: cluster IP" -m tcp --dport 8080 -j KUBE-SVC-KD42LXGMAX7AVWBX
    06 -A KUBE-SERVICES -d 10.254.0.1/32 -p tcp -m comment --comment
"default/kubernetes:https cluster IP" -m tcp --dport 443 -j
KUBE-SVC-NPX46M4PTMTKRN6Y
    07 -A KUBE-SERVICES -m comment --comment "kubernetes service nodeports; NOTE:
this must be the last rule in this chain" -m addrtype --dst-type LOCAL -j
KUBE-NODEPORTS
```

```
    08  -A KUBE-SVC-KD42LXGMAX7AVWBX -m comment --comment "default/tomcat:" -j
KUBE-SEP-TMJQS2NVDIZNCSFJ
    09  -A KUBE-SVC-NPX46M4PTMTKRN6Y -m comment --comment
"default/kubernetes:https" -m recent --rcheck --seconds 10800 --reap --name
KUBE-SEP-WSAW6HE4TRTZJXCF --mask 255.255.255.255 --rsource -j
KUBE-SEP-WSAW6HE4TRTZJXCF
    10  -A KUBE-SVC-NPX46M4PTMTKRN6Y -m comment --comment
"default/kubernetes:https" -j KUBE-SEP-WSAW6HE4TRTZJXCF
    11  COMMIT
```

其中，第 5 条规则表示目标地址为 10.254.12.13/32，并且目标端口为 8080 的数据包将被 KUBE-SVC-KD42LXGMAX7AVWBX 自定义链中的规则处理。而第 8 行定义了匹配 KUBE-SEP-TMJQS2NVDIZNCSFJ 自定义链中规则的数据包，将由 TMJQS2NVDIZNCSFJ 链中的规则处理。第 2 行定义了匹配 TMJQS2NVDIZNCSFJ 规则的数据包，将进行目标地址转换，转换目标为 172.17.1.3:8080，这个地址正是前面分配给名为 tomcat 的 Pod 的 IP 地址，以及 Tomcat 的服务端口。

从上面的分析可知，从服务到 Pod 之间的访问是调用 iptables 的规则实现的。

在任意节点上访问前面定义的服务，如下所示：

```
    [root@localhost ~]# curl 10.254.12.13:8080
    <!DOCTYPE html>
    <html lang="en">
        <head>
            <meta charset="UTF-8" />
            <title>Apache Tomcat/8.5.40</title>
            <link href="favicon.ico" rel="icon" type="image/x-icon" />
            <link href="favicon.ico" rel="shortcut icon" type="image/x-icon" />
            <link href="tomcat.css" rel="stylesheet" type="text/css" />
        </head>

        <body>
            <div id="wrapper">
                <div id="navigation" class="curved container">
                    <span id="nav-home"><a href="https://tomcat.apache.org/">
Home</a></span>
                    <span id="nav-hosts"><a href="/docs/">Documentation</a>
</span>
                    <span id="nav-config"><a href="/docs/config/">
Configuration</a></span>
                    <span id="nav-examples"><a href="/examples/">Examples</a>
</span>
                    <span id="nav-wiki"><a href="https://wiki.apache.org/tomcat/
FrontPage">Wiki</a></span>
```

```
                    <span id="nav-lists"><a href="https://tomcat.apache.org/
lists.html">Mailing Lists</a></span>
                    <span id="nav-help"><a href="https://tomcat.apache.org/
findhelp.html">Find Help</a></span>
                    <br class="separator" />
                </div>
                <div id="asf-box">
                    <h1>Apache Tomcat/8.5.40</h1>
                </div>
                <div id="upper" class="curved container">
                    <div id="congrats" class="curved container">
                        <h2>If you're seeing this, you've successfully installed
Tomcat. Congratulations!</h2>
                    </div>
    …
```

8.3　Flannel

Flannel 是一个专为 Kubernetes 定制的三层网络解决方案，主要用于解决容器的跨主机通信问题。本节将详细介绍 Flannel 的基本情况以及安装和使用方法。

8.3.1　Flannel 简介

Flannel 是一个 Kubernetes 网络插件，专门用于设置 Kubernetes 集群中的容器的网络地址空间。Flannel 利用 etcd 来存储整个集群的网络配置。例如，用户可以设置整个集群中所有容器的 IP 地址都取自网络 10.1.0.0/16。

在每个节点中，都运行着 Flannel 的代理服务 flanneld。该代理程序会为当前节点从集群的网络地址空间中，获取一个子网，本节点中所有的容器的 IP 地址都将从该子网中分配。所有的网络配置信息，都将存储在 etcd 中。

Flannel 提供了多种后端机制，例如 udp、vxlan 等。通过这些机制，实现了跨主机转发容器间的网络流量，完成容器间的跨主机通信。

图 8-6 描述了在 Flannel 网络中，容器之间的数据通信。首先，容器中的应用程序将数据包通过自己的网络接口 eth0 发送出去。然后，数据包会发送到虚拟网络接口 veth。而 veth 与虚拟网桥 docker0 桥接在一起，可以直接通信。因此，数据包通过 docker0 发送到虚拟网络接口 flannel0。而 Flannel 在 etcd 中存储了各个子网的路由规则，所以 flanneld 在查找路由规则之后，通过节点的网络接口 eth0 发送到其他的节点。数据包在到达目标节点后，在传输层交给 flanneld 守候进程处理。数据被解包，发送给 flannel0 虚拟网络接口。经过路由之后，发送给 docker0 网桥，再到达虚拟网络接口 veth，最后到达目标容器。

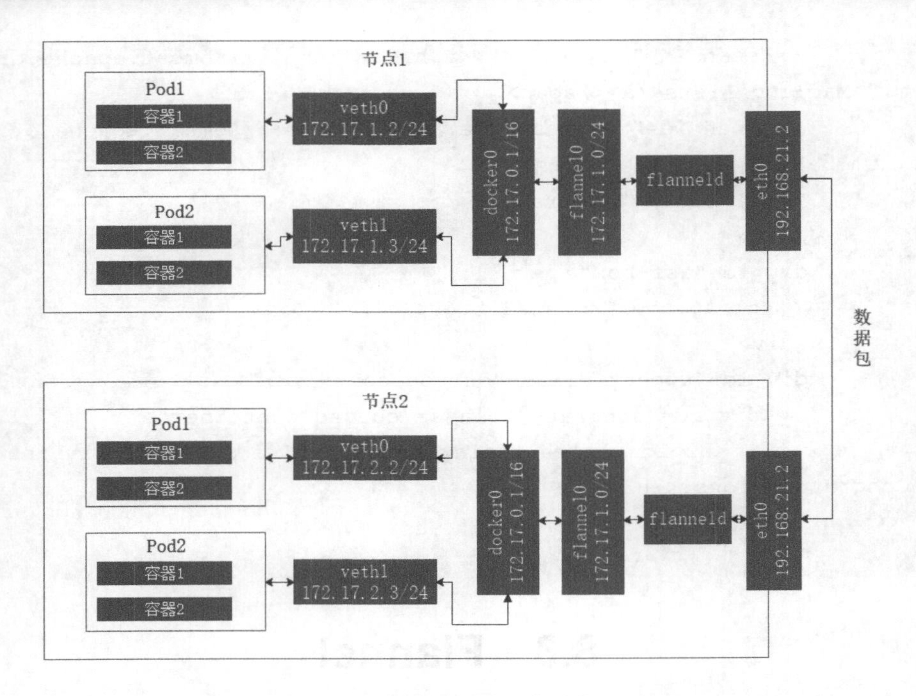

图 8-6 Flannel 跨节点通信

8.3.2 安装 Flannel

在 CentOS 中安装 Flannel 很简单，直接使用 yum 命令即可，如下所示：

```
[root@localhost ~]# yum -y install flannel
```

当然，用户也可以使用其他的安装方式，例如通过编译源代码或者下载二进制文件。对于初学者来说，通过操作系统的软件包管理工具来安装 Flannel 是非常容易掌握的。因为所有的节点都需要用到 Flannel，所以用户需要在每个节点上执行以上命令。

然后在 Master 节点上执行以下命令，在 etcd 中配置 Flannel 的网络信息：

```
    [root@localhost ~]# etcdctl --endpoints http://192.168.21.135:2379 set
/coreos.com/network/config '{"Network": "10.0.0.0/16", "SubnetLen": 24,
"SubnetMin": "10.0.1.0","SubnetMax": "10.0.20.0", "Backend": {"Type": "vxlan"}}'
    {"Network": "10.0.0.0/16", "SubnetLen": 24, "SubnetMin":
"10.0.1.0","SubnetMax": "10.0.20.0", "Backend": {"Type": "vxlan"}}
```

在上面的命令中，Network 用来指定 Flannel 所使用的网络 ID，后面分配给节点的子网都从该网络中分配。SubnetLen 用来指定分配给节点的虚拟网桥 docker0 的 IP 地址的子网掩码的长度。SubnetMin 用来指定最小子网的 ID。SudbnetMax 用来指定最大子网的 ID。在上面的命令中，我们指定最小子网为 10.0.1.0/24，最大子网为 10.0.20.0/24，由于每个节点分配一个子网，因此可以支持 20 个节点。Backend 用于指定数据包以什么方式转发，默认为 udp 模式，host-gw 模式性能最好，但不能跨宿主机网络。

然后修改 Flannel 的配置文件/etc/sysconfig/flanneld，增加 etcd 的访问地址，内容如下：

```
# Flanneld configuration options

# etcd url location.  Point this to the server where etcd runs
FLANNEL_ETCD_ENDPOINTS="http://192.168.21.135:2379,http://192.168.21.136:2379,http://192.168.21.137:2379"

# etcd config key.  This is the configuration key that flannel queries
# For address range assignment
FLANNEL_ETCD_PREFIX="/atomic.io/network"

# Any additional options that you want to pass
#FLANNEL_OPTIONS=""
```

设置完成，在每个节点上启动 Flannel，命令如下：

```
[root@localhost ~]# systemctl start flanneld
```

注　意

用户需要在启动 Flannel 前启动 Docker。

然后，用户可以在任意节点上执行以下命令，查看保存在 etcd 中的子网信息，如下所示：

```
[root@localhost ~]# etcdctl ls /atomic.io/network/subnets
/atomic.io/network/subnets/10.0.1.0-24
/atomic.io/network/subnets/10.0.7.0-24
/atomic.io/network/subnets/10.0.12.0-24
```

从上面的输出结果可知，目前已经为 3 个节点分配了子网，分别是 10.0.1.0/24、10.0.7.0/24 和 10.0.12.0/24。

在每个节点上查看网络接口，可以看到多出一个以 flannel 开头的虚拟网络接口，该网络接口的 IP 地址都分别位于 etcd 中对应节点的子网中。例如，下面是其中一个节点的网络接口情况：

```
[root@localhost ~]# ip address show
…
4: flannel.1: <BROADCAST,MULTICAST,UP,LOWER_UP> mtu 1450 qdisc noqueue state UNKNOWN group default
    link/ether 3e:0d:0d:13:ef:63 brd ff:ff:ff:ff:ff:ff
    inet 10.0.1.0/32 scope global flannel.1
       valid_lft forever preferred_lft forever
    inet6 fe80::3c0d:dff:fe13:ef63/64 scope link
       valid_lft forever preferred_lft forever
```

到此为止，Flannel 已经安装成功了。接下来需要在各个节点上面配置 Docker，修改其启动参数，使其能够从 Flannel 分配给当前节点的子网中获取 IP 地址。

在 Flannel 启动之后，会生成一个环境变量文件，包含了当前主机要使 Flannel 通信的相关参数，如下所示：

```
[root@localhost ~]# cat /run/flannel/subnet.env
FLANNEL_NETWORK=10.0.0.0/16
FLANNEL_SUBNET=10.0.1.1/24
FLANNEL_MTU=1450
FLANNEL_IPMASQ=false
```

用户可以使用以下命令将其转换为 Docker 的启动参数：

```
[root@localhost ~]# /usr/libexec/flannel/mk-docker-opts.sh
```

默认情况下，生成的 Docker 启动参数位于/run 目录中，其名称为 docker_opts.env，代码如下：

```
[root@localhost ~]# cat /run/docker_opts.env
DOCKER_OPT_BIP="--bip=10.0.1.1/24"
DOCKER_OPT_IPMASQ="--ip-masq=true"
DOCKER_OPT_MTU="--mtu=1450"
DOCKER_OPTS=" --bip=10.0.1.1/24 --ip-masq=true --mtu=1450"
```

修改 Docker 的服务单元文件/lib/systemd/system/docker.service，增加启动参数，如下所示：

```
#EnvironmentFile=-/etc/sysconfig/docker-network
EnvironmentFile=-/run/docker_opts.env
```

然后重启 Docker，命令如下：

```
[root@localhost ~]# systemctl daemon-reload
[root@localhost ~]# systemctl restart docker
```

再次查看 docker0 虚拟网桥的 IP 地址，就会发现其 IP 地址已经属于 flannel 分配给当前节点的子网，如下所示：

```
[root@localhost ~]# ip address show
…
3: docker0: <NO-CARRIER,BROADCAST,MULTICAST,UP> mtu 1500 qdisc noqueue state
DOWN group default
    link/ether 02:42:93:fe:a9:f1 brd ff:ff:ff:ff:ff:ff
    inet 10.0.1.1/24 scope global docker0
       valid_lft forever preferred_lft forever
4: flannel.1: <BROADCAST,MULTICAST,UP,LOWER_UP> mtu 1450 qdisc noqueue state
UNKNOWN group default
    link/ether 3e:0d:0d:13:ef:63 brd ff:ff:ff:ff:ff:ff
    inet 10.0.1.0/32 scope global flannel.1
       valid_lft forever preferred_lft forever
    inet6 fe80::3c0d:dff:fe13:ef63/64 scope link
       valid_lft forever preferred_lft forever
```

最后，我们通过创建两个 Pod 来验证其网络的连通性。这两个 Pod 的 YAML 配置文件如下：

```
[root@localhost ~]# cat centos.yaml
apiVersion: v1
kind: Pod
metadata:
  name: centos
spec:
  containers:
  - name: centos
    image: centos
    command: [ "/bin/bash", "-c", "--" ]
    args: [ "while true; do sleep 30; done;" ]
[root@localhost ~]# cat centos1.yaml
apiVersion: v1
kind: Pod
metadata:
  name: centos1
spec:
  containers:
  - name: centos
    image: centos
    command: [ "/bin/bash", "-c", "--" ]
    args: [ "while true; do sleep 30; done;" ]
```

然后分别使用以下命令创建 Pod：

```
[root@localhost ~]# kubectl create -f centos.yaml
[root@localhost ~]# kubectl create -f centos1.yaml
```

查看所创建的 Pod 状态信息：

```
[root@localhost ~]# kubectl get pod -o wide
NAME     READY    STATUS  RESTARTS    AGE  IP           NODE
centos   1/1      Running     0         3m     10.0.12.2  192.168.21.136
centos1      1/1      Running     0         1m     10.0.1.3  192.168.21.135
```

从上面的输出结果可知，这 2 个 Pod 分别在 192.168.21.135 和 192.168.21.136 这两个节点上，其 IP 地址分别为 10.0.1.3 和 10.0.12.2。

先进入名为 centos1 的 Pod：

```
[root@localhost ~]# kubectl exec -it centos1 -- /bin/sh
```

查看其 IP 地址，如下所示：

```
sh-4.2# ip address show
    1: lo: <LOOPBACK,UP,LOWER_UP> mtu 65536 qdisc noqueue state UNKNOWN group
default qlen 1000
```

```
    link/loopback 00:00:00:00:00:00 brd 00:00:00:00:00:00
    inet 127.0.0.1/8 scope host lo
      valid_lft forever preferred_lft forever
    inet6 ::1/128 scope host
      valid_lft forever preferred_lft forever
7: eth0@if8: <BROADCAST,MULTICAST,UP,LOWER_UP> mtu 1450 qdisc noqueue state
UP group default
    link/ether 02:42:0a:00:01:03 brd ff:ff:ff:ff:ff:ff link-netnsid 0
    inet 10.0.1.3/24 scope global eth0
      valid_lft forever preferred_lft forever
    inet6 fe80::42:aff:fe00:103/64 scope link
      valid_lft forever preferred_lft forever
```

通过 ping 命令测试与另外一个 Pod 的连通性，如下所示：

```
sh-4.2# ping 10.0.12.2
PING 10.0.12.2 (10.0.12.2) 56(84) bytes of data.
64 bytes from 10.0.12.2: icmp_seq=1 ttl=62 time=0.743 ms
64 bytes from 10.0.12.2: icmp_seq=2 ttl=62 time=2.20 ms
64 bytes from 10.0.12.2: icmp_seq=3 ttl=62 time=1.11 ms
64 bytes from 10.0.12.2: icmp_seq=4 ttl=62 time=1.36 ms
…
```

从上面的输出结果可知，尽管这 2 个 Pod 分别位于不同的节点，但是它们之间可以直接访问。

第 9 章

Kubernetes Dashboard

Kubernetes 不仅提供强大的命令行管理工具，还提供了基于 Web 的图形化管理界面，这就是本章将要介绍的 Dashboard（仪表盘）。通过 Kubernetes Dashboard，管理员可以非常方便地管理集群。

本章涉及的知识点有：

- Kubernetes Dashboard 配置文件：主要介绍官方提供的 Kubernetes Dashboard 配置文件及其内容。
- 安装 Dashboard：介绍 Kubernetes Dashboard 的安装方法。
- Dashboard 使用方法：主要介绍如何使用 Kubernetes Dashboard 管理集群。

9.1 Kubernetes Dashboard 配置文件

Kubernetes 官方为 Dashboard 专门提供了一个 YAML 配置文件，用来给用户部署 Dashboard。本节将对这个官方的 YAML 配置进行详细介绍，以便于后面的 Dashboard 部署。

9.1.1 Kubernetes 角色控制

Kubernetes 提供了基于角色的安全控制以及授权机制。其中，涉及了几个非常重要的概念，例如用户、角色、角色绑定以及 Secret 等，下面分别进行介绍。

1. 用户

Kubernetes 提供了两种用户，分别为 User 和 ServiceAccount。其中 User 是给管理员使用的，而 ServiceAccount 则是为 Pod 中的进程提供身份信息。当 Pod 的进程访问 API Server 时，它们会与一个 ServiceAccount，即服务账号相关联，通过该服务账号来判断该进程是否有权限访问 API Server。

当用户在创建 Pod 时，如果没有指定 ServiceAccount，则系统会自动在与该 Pod 所在的命名空间下，为其指定一个特殊的 ServiceAccount，其名称为 default。

管理员可以通过以下命令查看当前系统中的 ServiceAccount，如下所示：

```
[root@localhost ~]# kubectl get sa --all-namespaces
NAMESPACE       NAME        SECRETS        AGE
default              default          0              1d
kube-system          default          0              1d
```

从上面的输出结果可知，当前集群中存在 2 个名为 default 的 ServiceAccount，分别属于 default 和 kube-system 这 2 个命名空间。

2．角色

与其他的系统一样，在 Kubernetes 中，角色也是代表一系列权限的集合，权限以纯粹的累加形式累计。Kubernetes 支持 2 种角色，分别为 Role 和 ClusterRole。其中 Role 在命名空间内有效，不可以跨命名空间；ClusterRole 则是在整个集群范围内都有效，可以跨命名空间。

例如，下面的代码定义了一个 Role：

```
kind: Role
apiVersion: rbac.authorization.k8s.io/ v1alpha1
metadata:
  namespace: default
  name: pod-reader
rules:
- apiGroups: [""]
  resources: ["pods"]
  verbs: ["get", "watch", "list"]
```

可以得知，上面的 Role 可以访问的资源为 pods，可以使用的操作有 get、watch 和 list。

ClusterRole 除了可以授予与 Role 相同的权限之外，还可以授权集群范围内的资源、非资源类型的对象，以及跨命名空间的资源的访问权限。

3．角色绑定

角色绑定的作用是将角色映射到用户，从而让这些用户拥有该角色的权限。与 2 种角色相对应，角色绑定也分为 2 种，即 RoleBinding 和 ClusterRoleBinding。

4．Secret

关于 Secret，前面已经介绍过了。所谓 Secret，是一个包含少量敏感信息如密码、令牌或秘钥的对象。把这些信息保存在 Secret 对象中，可以在使用这些信息时加以控制，并可以降低信息泄露的风险。

9.1.2 kubernetes-dashboard.yaml

Kubernetes 官方为 Dashboard 提供了一个标准的 YAML 配置文件，其网址为：https://raw.githubusercontent.com/kubernetes/dashboard/master/aio/deploy/recommended/kubernetes-dashboard.yaml。

我们将该文件下载下来，分析一下它的代码。

```
01 # Copyright 2017 The Kubernetes Authors.
02 #
03 # Licensed under the Apache License, Version 2.0 (the "License");
04 # you may not use this file except in compliance with the License.
05 # You may obtain a copy of the License at
06 #
07 #    http://www.apache.org/licenses/LICENSE-2.0
08 #
09 # Unless required by applicable law or agreed to in writing, software
10 # distributed under the License is distributed on an "AS IS" BASIS,
11 # WITHOUT WARRANTIES OR CONDITIONS OF ANY KIND, either express or implied.
12 # See the License for the specific language governing permissions and
13 # limitations under the License.
14
15 # ------------------- Dashboard Secrets ------------------- #
16
17 apiVersion: v1
18 kind: Secret
19 metadata:
20   labels:
21     k8s-app: kubernetes-dashboard
22   name: kubernetes-dashboard-certs
23   namespace: kube-system
24 type: Opaque
25
26 ---
27
28 apiVersion: v1
29 kind: Secret
30 metadata:
31   labels:
32     k8s-app: kubernetes-dashboard
33   name: kubernetes-dashboard-csrf
34   namespace: kube-system
35 type: Opaque
36 data:
37   csrf: ""
38
39 ---
40 # ------------------- Dashboard Service Account ------------------- #
41
42 apiVersion: v1
43 kind: ServiceAccount
```

```
44 metadata:
45   labels:
46     k8s-app: kubernetes-dashboard
47   name: kubernetes-dashboard
48   namespace: kube-system
49
50 ---
51 # ------------------- Dashboard Role & Role Binding ------------------- #
52
53 kind: Role
54 apiVersion: rbac.authorization.k8s.io/ v1alpha1
55 metadata:
56   name: kubernetes-dashboard-minimal
57   namespace: kube-system
58 rules:
59   # Allow Dashboard to create 'kubernetes-dashboard-key-holder' secret.
60 - apiGroups: [""]
61   resources: ["secrets"]
62   verbs: ["create"]
63   # Allow Dashboard to create 'kubernetes-dashboard-settings' config map.
64 - apiGroups: [""]
65   resources: ["configmaps"]
66   verbs: ["create"]
67   # Allow Dashboard to get, update and delete Dashboard exclusive secrets.
68 - apiGroups: [""]
69   resources: ["secrets"]
70   resourceNames: ["kubernetes-dashboard-key-holder",
"kubernetes-dashboard-certs", "kubernetes-dashboard-csrf"]
71   verbs: ["get", "update", "delete"]
72   # Allow Dashboard to get and update 'kubernetes-dashboard-settings'
config map.
73 - apiGroups: [""]
74   resources: ["configmaps"]
75   resourceNames: ["kubernetes-dashboard-settings"]
76   verbs: ["get", "update"]
77   # Allow Dashboard to get metrics from heapster.
78 - apiGroups: [""]
79   resources: ["services"]
80   resourceNames: ["heapster"]
81   verbs: ["proxy"]
82 - apiGroups: [""]
83   resources: ["services/proxy"]
84   resourceNames: ["heapster", "http:heapster:", "https:heapster:"]
```

```
 85   verbs: ["get"]
 86
 87 ---
 88 apiVersion: rbac.authorization.k8s.io/ v1alpha1
 89 kind: RoleBinding
 90 metadata:
 91   name: kubernetes-dashboard-minimal
 92   namespace: kube-system
 93 roleRef:
 94   apiGroup: rbac.authorization.k8s.io
 95   kind: Role
 96   name: kubernetes-dashboard-minimal
 97 subjects:
 98 - kind: ServiceAccount
 99   name: kubernetes-dashboard
100   namespace: kube-system
101
102 ---
103 # ------------------- Dashboard Deployment ------------------- #
104
105 kind: Deployment
106 apiVersion: apps/v1
107 metadata:
108   labels:
109     k8s-app: kubernetes-dashboard
110   name: kubernetes-dashboard
111   namespace: kube-system
112 spec:
113   replicas: 1
114   revisionHistoryLimit: 10
115   selector:
116     matchLabels:
117       k8s-app: kubernetes-dashboard
118   template:
119     metadata:
120       labels:
121         k8s-app: kubernetes-dashboard
122     spec:
123       containers:
124       - name: kubernetes-dashboard
125         image: k8s.gcr.io/kubernetes-dashboard-amd64:v1.10.1
126         ports:
127         - containerPort: 8443
```

```
    128            protocol: TCP
    129          args:
    130            - --auto-generate-certificates
    131            # Uncomment the following line to manually specify Kubernetes
API server Host
    132            # If not specified, Dashboard will attempt to auto discover the
API server and connect
    133            # to it. Uncomment only if the default does not work.
    134            # - --apiserver-host=http://my-address:port
    135          volumeMounts:
    136          - name: kubernetes-dashboard-certs
    137            mountPath: /certs
    138            # Create on-disk volume to store exec logs
    139          - mountPath: /tmp
    140            name: tmp-volume
    141          livenessProbe:
    142            httpGet:
    143              scheme: HTTPS
    144              path: /
    145              port: 8443
    146            initialDelaySeconds: 30
    147            timeoutSeconds: 30
    148        volumes:
    149        - name: kubernetes-dashboard-certs
    150          secret:
    151            secretName: kubernetes-dashboard-certs
    152        - name: tmp-volume
    153          emptyDir: {}
    154        serviceAccountName: kubernetes-dashboard
    155        # Comment the following tolerations if Dashboard must not be deployed
on master
    156        tolerations:
    157        - key: node-role.kubernetes.io/master
    158          effect: NoSchedule
    159
    160  ---
    161  # ------------------- Dashboard Service ------------------- #
    162
    163  kind: Service
    164  apiVersion: v1
    165  metadata:
    166    labels:
    167      k8s-app: kubernetes-dashboard
```

```
168    name: kubernetes-dashboard
169    namespace: kube-system
170  spec:
171    ports:
172      - port: 443
173        targetPort: 8443
174    selector:
175      k8s-app: kubernetes-dashboard
```

从上面的代码可知，kubernetes-dashboard.yaml 文件分为 6 个部分，这 6 个部分分别为 Dashboard Secrets、Dashboard Service Account、Dashboard Role、Role Binding、Dashboard Deployment 和 Dashboard Service。

其中，第 17~37 行定义了 2 个 Secret，第 1 个 Secret 名为 kubernetes-dashboard-certs，第 2 个 Secret 名为 kubernetes-dashboard-csrf。这 2 个 Secret 的类型都为 Opaque，位于 kube-system 命名空间中。

第 42~48 行定义了名为 kubernetes-dashboard 的 ServiceAccount，即 Dashboard 的用户。

第 53~85 行定义了 Role，即 Dashboard 的角色信息，其角色名称为 kubernetes-dashboard-minimal，rules 字段中列出了其拥有的多个权限。通过名称我们可以猜到，这个权限级别是比较低的。

第 88~100 行定义了 RoleBinding，即 Dashboard 的角色绑定，其名称为 kubernetes-dashboard-minimal，roleRef 字段用来指定被绑定的角色，也叫 kubernetes-dashboard-minimal，即上面定义的角色，subjects 字段指定绑定的用户为 kubernetes-dashboard。

第 105~158 行定义了 Deployment。从上面第 154 行可知，Dashboard 的 Deployment 指定了其使用的 ServiceAccount 是 kubernetes-dashboard。此外，第 137 行将 Secret kubernetes-dashboard-certs 通过 volumes 挂载到 pod 内部的/certs 路径。由于在第 130 行指定了参数--auto-generate-certificates，因此 Dashboard 会自动生成证书。

第 163~175 行定义 Service，暴露了服务端口为 443，并且指定其目前端口为 8443，选择器为 k8s-app: kubernetes-dashboard。

可以发现，上面的代码定义了与 Dashboard 有关的各种资源。尽管上面的定义非常详细，但是由于网络原因，用户不可以直接通过上面的代码创建 Dashboard。

9.2　安装 Kubernetes Dashboard

Dashboard 是以 Pod 的形式提供服务的，所以安装 Dashboard 并不需要新的知识。本节将详细介绍 Dashboard 的安装方法。

9.2.1　官方安装方法

由于 Kubernetes 官方已经为 Dashboard 提供了一个 YAML 配置文件，该配置文件包含了 Dashboard 所需要的各种资源对象，因此用户可以直接使用以下命令快速部署 Dashboard：

```
[root@localhost ~]# kubectl apply -f https://raw.githubusercontent.com/
kubernetes/dashboard/master/aio/deploy/recommended/kubernetes-dashboard.yaml
```

在通过以上命令部署 Dashboard 的过程中，会从 k8s.gcr.io 拉取所需要的 kubernetes-dashboard-amd64 镜像文件。

9.2.2 自定义安装方法

尽管官方提供的安装方法非常简洁，但是在国内的网络环境中，用户却常常无法访问 k8s.gcr.io 去下载所需要的镜像，因此导致 Pod 创建失败。用户在查看 Dashboard 的 Pod 时，会发现它的状态为 ImagePullBackOff。

为了避免这个问题，用户可以通过国内的镜像服务器，预先将 kubernetes-dashboard 的镜像下载下来，然后再部署 Dashboard。这种国内的镜像服务器非常多，例如中国科学技术大学就提供了一个 k8s.gcr.io 的镜像服务器，用户可以通过以下命令下载镜像文件：

```
[root@localhost ~]# docker pull gcr.mirrors.ustc.edu.cn/google-containers/
kubernetes-dashboard-amd64:v2.1.0
```

下载完成之后，用户还需要设置镜像标签，如下所示：

```
[root@localhost ~]# docker tag gcr.mirrors.ustc.edu.cn/google-containers/
kubernetes-dashboard-amd64:v2.1.0 k8s.gcr.io/kubernetes-dashboard-amd64:v2.1.0
```

接下来就是编辑 YAML 配置文件。为了能够使读者更加清楚 Dashboard 的部署方法，我们在官方代码的基础上进行了简化和修改，去掉了与用户授权有关的内容，仅仅保留部署和服务这两部分，如下所示：

```
01 # ------------------- Dashboard Deployment ------------------- #
02
03 kind: Deployment
04 apiVersion: extensions/v1beta1
05 metadata:
06   labels:
07     k8s-app: kubernetes-dashboard
08   name: kubernetes-dashboard
09   namespace: kube-system
10 spec:
11   template:
12     metadata:
13       labels:
14         k8s-app: kubernetes-dashboard
15     spec:
16       containers:
17       - name: kubernetes-dashboard
18         image: k8s.gcr.io/kubernetes-dashboard-amd64:v2.1.0
19         resources:
```

```
20          limits:
21            cpu: 100m
22            memory: 50Mi
23          requests:
24            cpu: 100m
25            memory: 50Mi
26        ports:
27        - containerPort: 9090
28          protocol: TCP
29        args:
30          # - --auto-generate-certificates
31          # Uncomment the following line to manually specify Kubernetes API
server Host
32          # If not specified, Dashboard will attempt to auto discover the
API server and connect
33          # to it. Uncomment only if the default does not work.
34          - --apiserver-host=http://192.168.21.137:8080
35          - --heapster-host=http://heapster
36        livenessProbe:
37          httpGet:
38            scheme: HTTP
39            path: /
40            port: 9090
41          initialDelaySeconds: 30
42          timeoutSeconds: 30
43 ---
44 # ------------------- Dashboard Service ------------------- #
45
46 kind: Service
47 apiVersion: v1
48 metadata:
49   labels:
50     k8s-app: kubernetes-dashboard
51   name: kubernetes-dashboard
52   namespace: kube-system
53 spec:
54   ports:
55     - port: 9090
56       targetPort: 9090
57   selector:
58     k8s-app: kubernetes-dashboard
59   type: NodePort
```

此外，在上面的代码中，将 Dashboard 的访问方式由 HTTPS 修改为 HTTP，去掉了与证书有关的代码。第 27 行将服务端口修改为 9090。第 38 行将模式改为 HTTP。第 59 行将服务类型修改为 NodePort，以便于测试。

将以上代码保存为 kubernetes-dashboard.yaml，然后通过以下命令进行部署：

```
[root@localhost ~]# kubectl apply -f kubernetes-dashboard.yaml
```

查看 Pod 的状态，如下所示：

```
[root@localhost ~]# kubectl get pod -o wide --all-namespaces
NAMESPACE      NAME              READY          STATUS  RESTARTS    AGE
IP             NODE
kube-system     kubernetes-dashboard-1724149408-97377       1/1
Running 0          2m      172.17.0.2      127.0.0.1
```

从上面的输出结果可知，刚才创建的 Pod 已经处于运行状态了。

查询该 Pod 的日志，可以发现该 Pod 已经开始监听 9090 端口了：

```
[root@localhost ~]# kubectl log kubernetes-dashboard-1724149408-97377
--namespace=kube-system
W0422 05:13:39.677284    1613 cmd.go:337] log is DEPRECATED and will be removed
in a future version. Use logs instead.
2019/04/21 20:27:45 Starting overwatch
2019/04/21 20:27:45 Using apiserver-host location:
http://192.168.21.137:8080
2019/04/21 20:27:45 Skipping in-cluster config
2019/04/21 20:27:45 Using random key for csrf signing
2019/04/21 20:27:45 No request provided. Skipping authorization
2019/04/21 20:27:45 Successful initial request to the apiserver, version:
v1.5.2
2019/04/21 20:27:45 Generating JWE encryption key
2019/04/21 20:27:45 New synchronizer has been registered:
kubernetes-dashboard-key-holder-kube-system. Starting
2019/04/21 20:27:45 Starting secret synchronizer for
kubernetes-dashboard-key-holder in namespace kube-system
2019/04/21 20:27:51 Initializing JWE encryption key from synchronized object
2019/04/21 20:27:51 Creating remote Heapster client for http://heapster
2019/04/21 20:27:51 Serving insecurely on HTTP port: 9090
…
```

查看所创建的服务的状态，如下所示：

```
[root@localhost ~]# kubectl get svc --all-namespaces
NAMESPACE      NAME                      CLUSTER-IP  EXTERNAL-IP
PORT(S)        AGE
```

```
default        kubernetes              10.254.0.1      <none>
443/TCP        5h
kube-system    kubernetes-dashboard    10.254.175.41   <nodes>
9090:30963/TCP 49m
```

从上面的输出结果可知，服务的 ClusterIP 为 10.254.175.41，服务所监听的 9090 端口被映射到节点的 30963 端口。

继续查看服务的详细信息，可以发现，该服务所对应的 Endpoints 为 172.17.0.2:9090 正是我们前面创建的 Pod 的服务端口。

```
[root@localhost ~]# kubectl describe service kubernetes-dashboard
--namespace=kube-system
Name:                     kubernetes-dashboard
Namespace:                kube-system
Labels:                   k8s-app=kubernetes-dashboard
Selector:                 k8s-app=kubernetes-dashboard
Type:                     NodePort
IP:                       10.254.175.41
Port:                     <unset> 9090/TCP
NodePort:                 <unset> 30963/TCP
Endpoints:                172.17.0.2:9090
Session Affinity:         None
No events.
```

9.3 Dashboard 使用方法

Dashboard 为管理员提供了一个可视化的，基于浏览器的管理界面。通过 Dashboard，可以完成绝大部分的集群管理工作，从而可以在一定程度上代替 kubectl 命令。本节将详细介绍 Dashboard 的使用方法。

9.3.1 Dashboard 概况

在成功部署 Dashboard 之后，用户就可以通过节点 IP 加上端口来访问 Dashboard。例如，在上面的例子中，NodePort 为 30963，所以用户可以通过浏览器访问以下网址：

http://192.168.21.137:30963

Dashboard 的主界面如图 9-1 所示。

主界面左侧为菜单栏，包括集群、概况、工作负载、服务发现与负载均衡以及配置与存储等功能模块。每个功能模块下面又有多个子模块，例如，集群菜单下面就包括了命名空间、节点、持久化存储卷、角色以及存储类等管理模块。

主界面的右上角为创建按钮，用户可以通过该按钮创建各种资源，如图 9-2 所示。

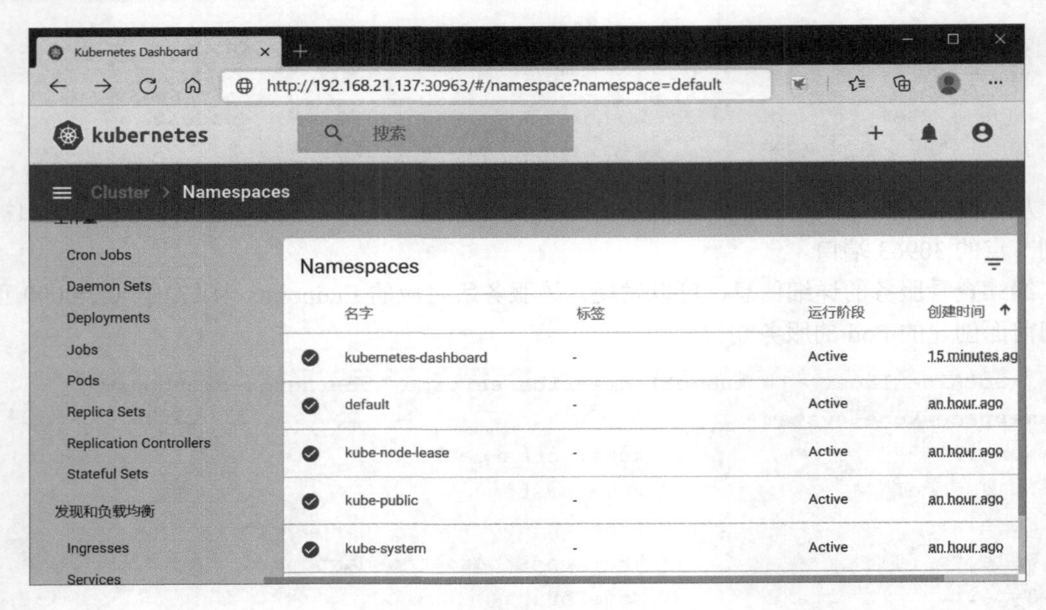

图 9-1　Dashboard 主界面

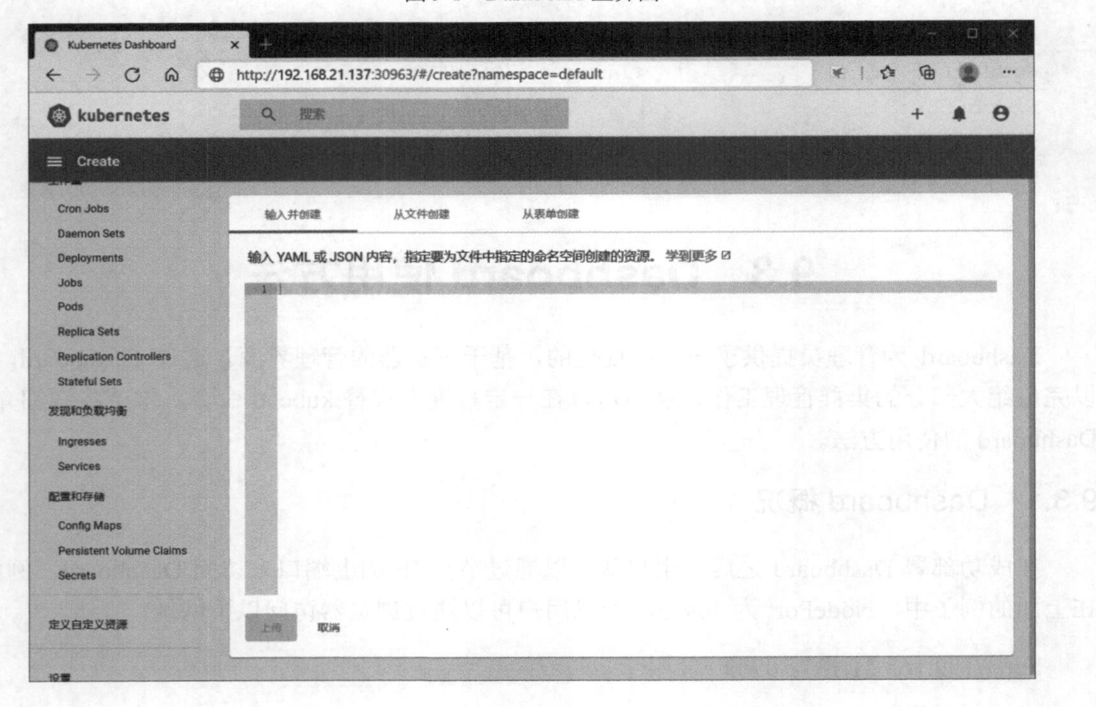

图 9-2　创建资源

创建资源界面包括 3 个标签页，分别为从文本输入框创建、从文件创建以及创建应用。其中，从文本输入框创建提供了一个多行文本框，用户可以在该文本框中输入 YAML 或者 JSON 配置文件。然后点击底部的上传按钮即可。从文件创建标签页提供了一个文件上传按钮，用户可以直接在本地编辑好 YAML 或者 JSON 配置文件，然后上传上去即可。创建应用标签页为用户提供了一个创建应用系统的便捷方式。

9.3.2　通过 Dashboard 创建资源

接下来，介绍一下如何通过 Dashboard 创建 Kubernetes 资源对象。点击图 9-2 所示窗口右上角的创建按钮，选择从文本输入框创建标签页，在文本框中输入以下代码：

```
apiVersion: extensions/v1beta1
kind: Deployment
metadata:
  name: nginx-deployment
spec:
  replicas: 3
  template:
  metadata:
    labels:
      app: nginx
        track: stable
    spec:
        containers:
    - name: nginx
      image: nginx:1.7.9
      ports:
      - containerPort: 80
```

点击底部的“上传”按钮，即可完成创建操作。

选择左侧的工作负载中的容器组菜单项，可以查看到刚才创建的 3 个 Pod，如图 9-3 所示。

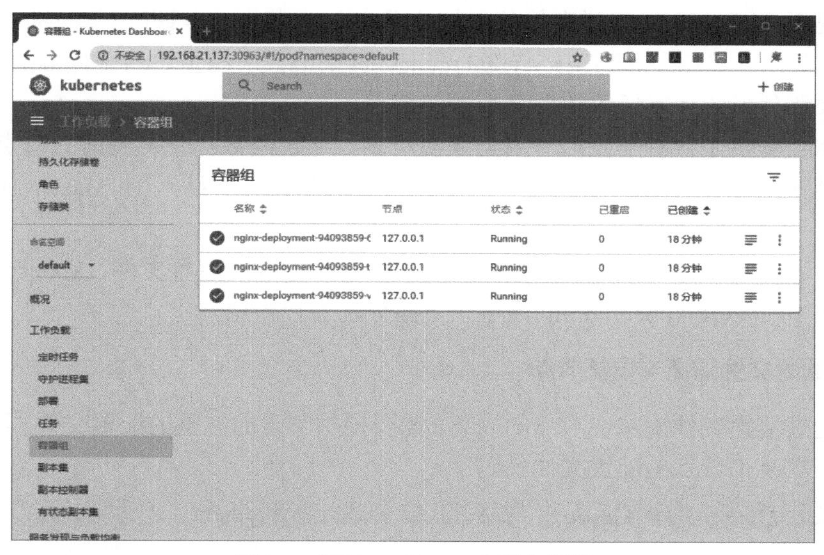

图 9-3　通过 Dashboard 查看容器组

通过 Dashboard 创建其他的资源对象的方法与上面的操作基本相同，读者可以自己去尝试，这里不再详细介绍。

第10章 Kubernetes集群管理

为了能够使得 Kubernetes 集群中的容器运行得更加稳定，用户需要熟练掌握 Kubernetes 集群的常用管理方法。例如，节点（Node）的管理、资源标签的管理以及监控方法等。本章将系统介绍这些方面的内容，以提高用户运维 Kubernetes 集群的水平。

10.1 管理节点

节点（Node）是 Kubernetes 集群中各项应用服务的实际提供者。因此，节点能否正常运行将直接影响到整个 Kubernetes 集群的服务质量。本节将详细介绍在 Kubernetes 集群运维过程中经常用到的节点管理方法。

10.1.1 节点的隔离与恢复

通常情况下，Kubernetes 集群中的节点都是由 Master 节点统一调度、管理的。在 Kubernetes 集群运行的过程中，难免会遇到某些特殊的情况，例如节点所在主机的硬件维护或者升级。此时，用户需要避免新的计算任务被调度到当前节点。

Kubernetes 提供了一种节点的隔离与恢复机制，通过隔离，使得当前的节点脱离 Master 节点的调度范围，在硬件维护完成之后，通过恢复机制又重新纳入到 Master 节点的调度范围中。

用户可以通过几种方式来实现节点的隔离和恢复，分别为配置文件、kubectl patch 命令、kubectl cordon 命令以及 kubectl drain 命令。下面分别进行介绍。

1. 通过配置文件隔离与恢复节点

如果是通过配置文件来实现节点的隔离，需要提供节点的完整的配置文件。用户可以通过以下命令来获取某个节点的配置文件代码：

```
[root@localhost ~]# kubectl get node node1 -o yaml
apiVersion: v1
kind: Node
metadata:
  annotations:
```

```
    kubeadm.alpha.kubernetes.io/cri-socket: /var/run/dockershim.sock
    node.alpha.kubernetes.io/ttl: "0"
    volumes.kubernetes.io/controller-managed-attach-detach: "true"
  creationTimestamp: "2020-11-29T02:58:46Z"
  labels:
    beta.kubernetes.io/arch: amd64
    beta.kubernetes.io/os: linux
    kubernetes.io/arch: amd64
    kubernetes.io/hostname: node1
    kubernetes.io/os: linux
  managedFields:
  - apiVersion: v1
    fieldsType: FieldsV1
    fieldsV1:
      f:metadata:
…
```

其中 node1 为节点的名称，-o 选项用来指定输出格式为 YAML。将以上代码保存为 unschedulable.yaml 文件，然后在其中的 spec 部分增加以下代码：

```
unschedulable: true
```

修改后的 unschedulable.yaml 文件的内容如下：

```
01 apiVersion: v1
02 kind: Node
03 metadata:
04   annotations:
05     flannel.alpha.coreos.com/backend-data:
'{"VNI":1,"VtepMAC":"1a:c9:0b:f8:a4:5f"}'
06     flannel.alpha.coreos.com/backend-type: vxlan
07     flannel.alpha.coreos.com/kube-subnet-manager: "true"
08     flannel.alpha.coreos.com/public-ip: 192.168.2.120
09     kubeadm.alpha.kubernetes.io/cri-socket: /var/run/dockershim.sock
10     node.alpha.kubernetes.io/ttl: "0"
11     volumes.kubernetes.io/controller-managed-attach-detach: "true"
12 ...
13 spec:
14   unschedulable: true
15   podCIDR: 10.244.1.0/24
16   podCIDRs:
17   - 10.244.1.0/24
18 status:
19 ...
```

然后使用 kubectl replace 命令对节点状态进行修改，如下所示：

```
[root@localhost ~]# kubectl replace -f unschedulable.yaml
node/node1 replaced
```

当以上命令成功执行以后，查看一下 Node 的状态，命令如下：

```
[root@localhost ~]# kubectl get nodes
NAME                    STATUS                     ROLES    AGE   VERSION
localhost.localdomain   Ready                      master   34m   v1.19.4
node1               Ready,SchedulingDisabled   <none>       14m   v1.19.4
```

从上面的输出结果可知，名为 node1 的节点的状态列中增加了一个 SchedulingDisabled 状态项，表示当前的节点为隔离状态。查看 node1 的详细信息，也可以发现当前节点的 Unschedulable 的值为 true，如下所示：

```
[root@localhost ~]# kubectl describe node node1
Name:               node1
Roles:              <none>
Labels:             beta.kubernetes.io/arch=amd64
                    beta.kubernetes.io/os=linux
                    kubernetes.io/arch=amd64
                    kubernetes.io/hostname=node1
                    kubernetes.io/os=linux
Annotations:         kubeadm.alpha.kubernetes.io/cri-socket: /var/run/
dockershim.sock
                   node.alpha.kubernetes.io/ttl: 0
                   volumes.kubernetes.io/controller-managed-attach-detach:
true
CreationTimestamp:  Sun, 29 Nov 2020 10:58:46 +0800
Taints:             node.kubernetes.io/unschedulable:NoSchedule
Unschedulable:      true
Lease:
  HolderIdentity:  node1
  AcquireTime:     <unset>
  RenewTime:       Sun, 29 Nov 2020 11:07:15 +0800
…
```

当用户在 Kubernetes 集群中创建新的 Pod 时，Kubernetes 不会向处于 SchedulingDisabled 状态的节点进行调度，即 Kubernetes 不会在处于 SchedulingDisabled 状态的节点上创建新的 Pod。对于已经运行在该节点上的 Pod，则不会自动停止，用户需要手动停止这些 Pod，然后再进行维护。

在节点维护工作完成之后，用户想要恢复该节点，将其恢复到 Kubernetes 集群的管理之下，那么只需要将前面的 unschedulable.yaml 配置文件中的 unschedulable 修改为 false，其余代码无需修改，如下所示：

```
…
spec:
  unschedulable: false
  podCIDR: 10.244.1.0/24
  podCIDRs:
  - 10.244.1.0/24
status:
…
```

将修改后的代码保存为 schedulable.yaml，然后使用 kubectl replace 命令将其恢复，如下所示：

```
[root@localhost ~]# kubectl replace -f schedulable.yaml
node/node1 replaced
```

命令执行成功之后，查看节点的状态，如下所示：

```
[root@localhost ~]# kubectl get nodes
NAME                     STATUS    ROLES     AGE    VERSION
localhost.localdomain    Ready     master    66m    v1.19.4
node1                    Ready     <none>    46m    v1.19.4
```

从上面的输出结果可知，node1 的状态中的 SchedulingDisabled 已经被去掉，此时 node1 处于 Kubernetes 的调度范围之中。

注　意

如果在执行 kubectl replace 的时候出现以下错误：

```
Error from server (Conflict): error when replacing "schedulable.yaml": Operation cannot be fulfilled on nodes "node1": the object has been modified; please apply your changes to the latest version and try again
```

则表示资源的版本发生冲突，用户可以修改 YAML 配置文件，将其中的 resourceVersion 一行删除，如下所示：

```
resourceVersion: "11128"
```

然后再执行 kubectl replace 命令。

2. 通过 kubectl patch 命令隔离与恢复节点

kubectl patch 命令可以对 Kubernetes 集群中的资源进行更新，支持 JSON 和 YAML 两种格式。例如，下面的命令对名为 node1 的节点的配置进行修改，将其改为隔离状态：

```
[root@localhost ~]# kubectl patch node node1 -p
'{"spec":{"unschedulable":true}}'
node/node1 patched
```

在上面的命令中，-p 选项用来指定要应用的修改内容，此处使用 JSON 格式表示。其功能是将 node1 的配置文件中的 spec 主键下面的 unschedulable 主键的值修改为 true。当以上命令执行成功之后，查看 node1 的状态，如下所示：

```
[root@localhost ~]# kubectl get nodes
NAME                    STATUS                       ROLES     AGE     VERSION
localhost.localdomain   Ready                        master    3h20m   v1.19.4
node1                   Ready,SchedulingDisabled     <none>    3h      v1.19.4
```

从上面的输出结果可知，node1 的状态中已经增加了 SchedulingDisabled 状态。

如果想要恢复 node1，则只要将命令中的 unschedulable 主键的值修改为 false 即可，如下所示：

```
[root@localhost ~]# kubectl patch node node1 -p
'{"spec":{"unschedulable":false}}'
node/node1 patched
```

用户可以通过命令查看 node1 的状态。

3. 通过 kubectl cordon 命令隔离与恢复节点

前面介绍的 kubectl patch 不仅可以修改节点的状态，还可以修改 Kubernetes 中任何资源的配置。在新版本的 Kubernetes 中，kubectl cordon 命令专门用来对节点进行隔离和恢复。

kubectl cordon 命令的基本语法比较简单，只要提供节点的名称作为参数即可，如下所示：

```
[root@localhost ~]# kubectl cordon node1
node/node1 cordoned
[root@localhost ~]# kubectl get nodes
NAME                    STATUS                       ROLES     AGE     VERSION
localhost.localdomain   Ready                        master    3h31m   v1.19.4
node1                   Ready,SchedulingDisabled     <none>    3h11m   v1.19.4
```

如果想要恢复节点，则需要使用 kubectl uncordon 命令，如下所示：

```
[root@localhost ~]# kubectl uncordon node1
node/node1 uncordoned
[root@localhost ~]# kubectl get nodes
NAME                    STATUS     ROLES     AGE     VERSION
localhost.localdomain   Ready      master    3h32m   v1.19.4
node1                   Ready      <none>    3h12m   v1.19.4
```

4. 通过 kubectl drain 命令隔离与恢复节点

前面介绍的命令在隔离节点之后，被隔离的节点上的 Pod 并不会停止，也不会被驱逐。而 kubectl drain 命令除了隔离节点之外，还对被隔离的节点上的 Pod 进行驱逐。如下所示：

```
[root@localhost ~]# kubectl drain node1 --ignore-daemonsets
node/node1 cordoned
```

```
WARNING: ignoring DaemonSet-managed Pods: kube-system/kube-proxy-cckr7
node/node1 drained
[root@localhost ~]# kubectl get nodes
NAME                    STATUS                     ROLES    AGE    VERSION
localhost.localdomain   Ready                      master   3h38m  v1.19.4
node1                   Ready,SchedulingDisabled   <none>   3h18m  v1.19.4
```

对于使用 kubectl drain 命令隔离的节点，可以使用前面介绍的命令进行恢复，例如，下面调用 kubectl uncordon 命令恢复 node1：

```
[root@localhost ~]# kubectl uncordon node1
node/node1 uncordoned
```

10.1.2　节点的扩容

在 Kubernetes 集群运行一段时间之后，随着业务的增长，集群中的计算资源会出现不能满足需求的情况。此时，用户可以购买新的服务器，在上面部署 Kubernetes 的节点组件，并将其添加到 Kubernetes 集群中去，实现节点的扩容。下面详细介绍如何将一台新的 CentOS 8 的服务器加入到 Kubernetes 集群中。

（1）设置主机名。在加入到 Kubernetes 集群时，主机名会作为节点的名称，因此，新主机的主机名必须能够标识当前的主机，并且不能与 Kubernetes 集群中的其他节点的名称相同。

```
[root@localhost ~]# hostnamectl set-hostname node2
```

通过以上命令，将当前主机的主机名设置为 node2，表示这是当前 Kubernetes 集群中的第 2 个节点。

（2）关闭 firewalld 防火墙，命令如下：

```
[root@localhost ~]# systemctl disable firewalld.service
Removed /etc/systemd/system/multi-user.target.wants/firewalld.service.
Removed /etc/systemd/system/dbus-org.fedoraproject.FirewallD1.service.
[root@localhost ~]# systemctl stop firewalld.service
```

（3）关闭交换分区，命令如下：

```
[root@localhost ~]# swapoff -a && sysctl -w vm.swappiness=0
vm.swappiness = 0
```

修改/etc/fstab 文件，注释掉关于交换分区的挂载选项，如下所示：

```
#/dev/mapper/cl-swap     swap                    swap    defaults        0 0
```

（4）开启 iptable 的桥接功能，如下所示：

```
[root@localhost ~]# cat <<EOF | sudo tee /etc/sysctl.d/k8s.conf
> net.bridge.bridge-nf-call-ip6tables = 1
> net.bridge.bridge-nf-call-iptables = 1
> EOF
```

```
    sudo sysctl --systemnet.bridge.bridge-nf-call-ip6tables = 1
    net.bridge.bridge-nf-call-iptables = 1
    [root@localhost ~]# sudo sysctl --system
    * Applying /usr/lib/sysctl.d/10-default-yama-scope.conf ...
    kernel.yama.ptrace_scope = 0
    * Applying /usr/lib/sysctl.d/50-coredump.conf ...
    kernel.core_pattern =
|/usr/lib/systemd/systemd-coredump %P %u %g %s %t %c %h %e
    * Applying /usr/lib/sysctl.d/50-default.conf ...
    kernel.sysrq = 16
    kernel.core_uses_pid = 1
    kernel.kptr_restrict = 1
    net.ipv4.conf.all.rp_filter = 1
    net.ipv4.conf.all.accept_source_route = 0
    net.ipv4.conf.all.promote_secondaries = 1
    net.core.default_qdisc = fq_codel
    fs.protected_hardlinks = 1
    fs.protected_symlinks = 1
    * Applying /usr/lib/sysctl.d/50-libkcapi-optmem_max.conf ...
    net.core.optmem_max = 81920
    * Applying /usr/lib/sysctl.d/50-pid-max.conf ...
    kernel.pid_max = 4194304
    * Applying /etc/sysctl.d/99-sysctl.conf ...
    * Applying /etc/sysctl.d/k8s.conf ...
    * Applying /etc/sysctl.conf ...
```

（5）关闭 SELinux，命令如下：

```
    [root@localhost ~]# setenforce 0
    [root@localhost ~]# sed -i 's/^SELINUX=enforcing$/SELINUX=permissive/'
/etc/selinux/config
```

（6）配置 Kubernetes 的阿里云软件源，命令如下：

```
    [root@localhost ~]# cat <<EOF | sudo tee /etc/yum.repos.d/kubernetes.repo
    > [kubernetes]
    > name=Kubernetes
    > baseurl=http://mirrors.aliyun.com/kubernetes/yum/repos/kubernetes-el7-x86_64
    > enabled=1
    > gpgcheck=0
    > repo_gpgcheck=0
    > gpgkey=http://mirrors.aliyun.com/kubernetes/yum/doc/yum-key.gpg
    >       http://mirrors.aliyun.com/kubernetes/yum/doc/rpm-package-key.gpg
    > exclude=kubelet kubeadm kubectl
    > EOF
```

```
[kubernetes]
name=Kubernetes
baseurl=http://mirrors.aliyun.com/kubernetes/yum/repos/kubernetes-el7-x86
_64
enabled=1
gpgcheck=0
repo_gpgcheck=0
gpgkey=http://mirrors.aliyun.com/kubernetes/yum/doc/yum-key.gpg
      http://mirrors.aliyun.com/kubernetes/yum/doc/rpm-package-key.gpg
exclude=kubelet kubeadm kubectl
```

（7）配置 Docker 的阿里云软件源，命令如下：

```
[root@localhost ~]# yum-config-manager --add-repo
http://mirrors.aliyun.com/docker-ce/linux/centos/docker-ce.repo
Repository AppStream is listed more than once in the configuration
Repository extras is listed more than once in the configuration
Repository PowerTools is listed more than once in the configuration
Repository centosplus is listed more than once in the configuration
Adding repo from:
http://mirrors.aliyun.com/docker-ce/linux/centos/docker-ce.repo
```

（8）安装 Docker，命令如下：

```
[root@localhost ~]# dnf install -y docker-ce
```

安装完成之后，启用并启动 Docker，如下所示：

```
[root@localhost ~]# systemctl enable docker
Created symlink /etc/systemd/system/multi-user.target.wants/docker.service
→ /usr/lib/systemd/system/docker.service.
[root@localhost ~]# systemctl start docker
```

注　意

如果在安装 Docker 的过程中出现以下错误：

```
Error:
 Problem: package docker-ce-3:19.03.13-3.el7.x86_64 requires containerd.io >=
1.2.2-3, but none of the providers can be installed
```

则表示当前的系统中的 containerd.io 组件的版本过低，用户可以使用以下命令安装高版本的 containerd.io：

```
[root@localhost ~]# dnf install -y https://download.docker.com/linux/centos/8/
x86_64/edge/Packages/containerd.io-1.3.7-3.1.el8.x86_64.rpm
```

（9）安装 Kubernetes 组件，命令如下：

```
[root@localhost ~]# dnf install -y kubelet kubeadm --disableexcludes=
kubernetes
```

安装完成之后，启动 kubelet 服务，命令如下：

```
[root@localhost ~]# systemctl enable --now kubelet
Created symlink /etc/systemd/system/multi-user.target.wants/
kubelet.service → /usr/lib/systemd/system/kubelet.service.
[root@localhost ~]# systemctl restart kubelet
```

（10）拉取镜像，命令如下：

```
[root@localhost ~]# docker pull registry.aliyuncs.com/google_containers/
kube-proxy:v1.19.4
```

在节点主机上，用户需要安装 kube-proxy 组件，因此在上面的命令中，只需拉取 kube-proxy 的镜像文件。

（11）将新的节点注册到 Kubernetes 集群中，然后执行以下命令：

```
[root@localhost ~]# kubeadm join 192.168.2.119:6443 --token
gxyn1r.aovqrj7cj9au792p     --discovery-token-ca-cert-hash
sha256:ae5c96f119136a836228b14498f7a6cedb84a76e8e21ef262ee67a33e614c5a0
[preflight] Running pre-flight checks
        [WARNING IsDockerSystemdCheck]: detected "cgroupfs" as the Docker
cgroup driver. The recommended driver is "systemd". Please follow the guide at
https://kubernetes.io/docs/setup/cri/
        [WARNING FileExisting-tc]: tc not found in system path
        [WARNING Hostname]: hostname "node2" could not be reached
        [WARNING Hostname]: hostname "node2": lookup node2 on 192.168.2.1:53:
no such host
[preflight] Reading configuration from the cluster...
[preflight] FYI: You can look at this config file with 'kubectl -n kube-system
get cm kubeadm-config -oyaml'
[kubelet-start] Writing kubelet configuration to file
"/var/lib/kubelet/config.yaml"
[kubelet-start] Writing kubelet environment file with flags to file
"/var/lib/kubelet/kubeadm-flags.env"
[kubelet-start] Starting the kubelet
[kubelet-start] Waiting for the kubelet to perform the TLS Bootstrap...

This node has joined the cluster:
* Certificate signing request was sent to apiserver and a response was received.
* The Kubelet was informed of the new secure connection details.

Run 'kubectl get nodes' on the control-plane to see this node join the cluster.
```

其中，192.168.2.119:6443 为 Kubernetes 的 Master 主机的 IP 地址以及服务端口。--token 用来指定认证所用的 Token，该 Token 在初始化集群的时候自动生成。如果忘记了 Token，用户可以在 Master 主机上通过以下命令查看：

```
[root@localhost ~]# kubeadm token list
TOKEN          TTL       EXPIRES              USAGES          DESCRIPTION          EXTRA GROUPS
gxyn1r.aovqrj7cj9au792p19h      2020-11-30T10:38:52+08:00
authentication,signing      The default bootstrap token generated by 'kubeadm
init'.  system:bootstrappers:kubeadm:default-node-token
```

在上面的输出结果中，TOKEN 列就是当前集群的认证 Token。

--discovery-token-ca-cert-hash 选项用来指定当前集群的 CA 证书的 Hash 值。该 Hash 值是在初始化集群时自动生成的。如果忘记了该 Hash 值，用户可以使用以下命令获取：

```
[root@localhost ~]# openssl x509 -pubkey -in /etc/kubernetes/pki/ca.crt | openssl
rsa -pubin -outform der 2>/dev/null | openssl dgst -sha256 -hex | sed 's/^.* //'
ae5c96f119136a836228b14498f7a6cedb84a76e8e21ef262ee67a33e614c5a0
```

其中，-in 选项指定 CA 证书的路径。该命令的返回值即为所需要的 Hash 值。

（12）查看节点状态。在 Master 节点上面查看节点的状态，命令如下：

```
[root@localhost ~]# kubectl get nodes
NAME                     STATUS    ROLES     AGE     VERSION
localhost.localdomain    Ready     master    5h1m    v1.19.4
node1                    Ready     <none>    4h41m   v1.19.4
node2                    Ready     <none>    30m     v1.19.4
```

从上面的输出结果可知，新加入的 node2 已经处于 Ready（就绪）状态。

10.2 管理资源对象标签

标签是 Kubernetes 中一个非常重要的概念，一个标签就是一个“键-值对”。标签的主要功能是对集群中的各种资源进行多维度的分组管理。本节将详细介绍标签的使用方法。

10.2.1 查看资源标签

用户可以通过多种方式来查看资源的标签属性。例如，在查看节点的时候，用户可以使用 --show-labels 选项显示节点的标签，如下所示：

```
[root@localhost ~]# kubectl get nodes --show-labels
NAME                    STATUS    ROLES     AGE     VERSION    LABELS
localhost.localdomain   Ready     master    5h30m   v1.19.4
beta.kubernetes.io/arch=amd64,beta.kubernetes.io/os=linux,kubernetes.io/a
rch=amd64,kubernetes.io/hostname=localhost.localdomain,kubernetes.io/os=linux
,node-role.kubernetes.io/master=
```

```
    node1                Ready      <none>     5h10m   v1.19.4
    beta.kubernetes.io/arch=amd64,beta.kubernetes.io/os=linux,kubernetes.io/a
rch=amd64,kubernetes.io/hostname=node1,kubernetes.io/os=linux
    node2                Ready      <none>     60m     v1.19.4
    beta.kubernetes.io/arch=amd64,beta.kubernetes.io/os=linux,kubernetes.io/a
rch=amd64,kubernetes.io/hostname=node2,kubernetes.io/os=linux
```

从上面的输出结果可知，一个资源可以拥有多个标签。例如，node1 标签就包含 beta.kubernetes.io/arch=amd64、beta.kubernetes.io/os=linux、kubernetes.io/arch=amd64、kubernetes.io/hostname=node1 以及 kubernetes.io/os=linux 等 5 个标签。每个标签都是一个“键-值对”，等号前面的为键（K），等号后面的为值（Value）。

在使用 kube describe 命令查看资源信息的时候，也可以显示标签，如下所示：

```
    [root@localhost ~]# kubectl describe node node1
    Name:               node1
    Roles:              <none>
    Labels:             beta.kubernetes.io/arch=amd64
                        beta.kubernetes.io/os=linux
                        kubernetes.io/arch=amd64
                        kubernetes.io/hostname=node1
                        kubernetes.io/os=linux
    Annotations:        kubeadm.alpha.kubernetes.io/cri-socket: /var/run/
dockershim.sock
                        node.alpha.kubernetes.io/ttl: 0
                        volumes.kubernetes.io/controller-managed-attach-detach:
true
    CreationTimestamp:  Sun, 29 Nov 2020 10:58:46 +0800
    Taints:             <none>
    Unschedulable:      false
    Lease:
      HolderIdentity:  node1
      AcquireTime:     <unset>
      RenewTime:       Sun, 29 Nov 2020 16:21:37 +0800
    Conditions:
    …
```

10.2.2 添加资源标签

用户可以使用 kubectl label 命令非常方便地为资源添加各种自定义标签。该命令的基本语法如下：

```
    kubectl label [options]
```

其中常用的选项有--all 和--overwrite 等。下面以具体的例子来说明该命令的使用方法。

下面的命令为名称为 kube-proxy-zc9q8 的 Pod 设置一个标签，用来标识该 Pod 的状态，如下所示：

```
    [root@localhost ~]# kubectl label pods kube-proxy-zc9q8 unhealthy=true -n
kube-system
    pod/kube-proxy-zc9q8 labeled
```

其中 pods 表示当前设置的资源类型为 Pod，kube-proxy-zc9q8 为资源的名称，unhealthy=true 为要添加的标签，-n 选项用来指定资源所在的命名空间。当所要设置的资源不在当前默认的命名空间中时，需要通过-n 选项指定。

设置成功之后，用户就可以使用该标签来对资源进行筛选，如下所示：

```
    [root@localhost ~]# kubectl get pods --all-namespaces -l unhealthy=true
--show-labels
    NAMESPACE       NAME                 READY   STATUS  RESTARTS    AGE
    LABELS
    kube-system     kube-proxy-zc9q8     1/1     Running 1           92m
    controller-revision-hash=7f9fdcd659,k8s-app=kube-proxy,pod-template-gener
ation=1,unhealthy=true
```

在上面的命令中，-all-namespaces 选项表示显示所有的命名空间的资源。-l 选项表示通过标签对资源进行筛选，随后紧跟的就是标签。

以下命令对当前命名空间中所有的 Pod 设置标签：

```
    [root@localhost ~]# kubectl label pods --all status=unhealthy
```

其中--all 选项表示某个特定的命名空间中所有的指定类型的资源。

10.2.3　修改资源标签

如果用户想要修改资源的标签，需要使用--overwrite 选项，否则，会出现标签值已经存在的错误，导致标签修改失败。例如，下面的命令将名为 kube-proxy-zc9q8 的 Pod 的 unhealthy 标签的值修改为 false：

```
    [root@localhost ~]# kubectl label pods kube-proxy-zc9q8 unhealthy=false -n
kube-system --overwrite
    pod/kube-proxy-zc9q8 labeled
```

10.2.4　删除资源标签

删除资源标签非常简单，只要在标签的后面加上一个“-”即可。例如，假设名为 node1 的节点有一个名为 role 的标签，通过以下命令可以将其删除：

```
    [root@localhost ~]# kubectl label node node1 role-
    node/node1 labeled
```

注　意

“-”符号要紧跟在键的后面，中间不能有空格。

10.3　管理命名空间

关于命名空间的基本概念，已经在第 1 章中介绍过了。在 Kubernetes 集群中，通过自定义命名空间，可以有效地对各种资源对象进行隔离管理。本节将详细介绍命名空间的使用方法。

10.3.1　创建命名空间

命名空间的创建方法与其他的资源对象创建方法基本相同。首先编写一个 YAML 配置文件，然后通过 kubectl create 命令进行创建。

假设当前集群需要分成两个相互独立的工作区，其中一个区域作为生产环境，用来运行一些正式的服务；另外一个区域作为测试环境，开发人员可以在里面任意地创建和删除各种资源。

为了实现这个需求，我们需要创建两个命名空间，其中一个名称为 development，另外一个名称为 production。

其中 development 命名空间的 YAML 配置文件的名称为 namespace-devel.yaml，其代码如下：

```
01 apiVersion: v1
02 kind: Namespace
03 metadata:
04   name: development
```

其中第 4 行通过 name 指定当前命名空间的名称为 development。

production 命名空间的 YAML 配置文件的名称为 namespace-prod.yaml，其代码如下：

```
01 apiVersion: v1
02 kind: Namespace
03 metadata:
04   name: production
```

然后使用以下命令创建以上 2 个命名空间：

```
[root@localhost ~]# kubectl create -f namespace-devel.yaml
namespace/development created
[root@localhost ~]# kubectl create -f namespace-production.yaml
namespace/production created
```

创建完成之后，可以使用以下命令查看是否创建成功：

```
[root@localhost ~]# kubectl get namespaces
NAME                        STATUS          AGE
```

```
default              Active        7h52m
development          Active        20s
kube-node-lease      Active        7h52m
kube-public          Active        7h52m
kube-system          Active        7h52m
production           Active        15s
```

从上面的输出结果可知，development 和 production 这 2 个命名空间已经出现在列表中。除了创建命名空间之外，用户还需要创建 2 个 Context，即上下文环境，这 2 个 Context 分别属于上面创建的 2 个命名空间。命令如下：

```
[root@localhost ~]# kubectl config set-context ctx-dev
--namespace=development --cluster=kubernetes --user=kubernetes-admin
Context "ctx-dev" created.
[root@localhost ~]# kubectl config set-context ctx-prod
--namespace=production --cluster=kubernetes --user=kubernetes-admin
Context "ctx-prod" created.
```

其中 ctx-dev 和 ctx-prod 分别为 2 个 Context 的名称。--namespace 选项用来指定 Context 所属的命名空间。--cluster 选项指定集群名称，--user 选项为 Context 所属的用户。为了简化操作，此处分别使用 Kubernetes 集群本身的名称和用户名。

创建完成之后，用户可以使用以下命令查看当前集群中的 Context：

```
[root@localhost ~]# kubectl config view
apiVersion: v1
clusters:
- cluster:
    certificate-authority-data: DATA+OMITTED
    server: https://192.168.2.119:6443
  name: kubernetes
contexts:
- context:
    cluster: kubernetes
    namespace: development
    user: kubernetes-admin
  name: ctx-dev
- context:
    cluster: kubernetes
    namespace: production
    user: kubernetes-admin
  name: ctx-prod
- context:
    cluster: kubernetes
    user: kubernetes-admin
  name: kubernetes-admin@kubernetes
```

```
current-context: kubernetes-admin@kubernetes
kind: Config
preferences: {}
users:
- name: kubernetes-admin
  user:
    client-certificate-data: REDACTED
    client-key-data: REDACTED
```

从上面的输出结果可知，当前集群中有 3 个 Context，分别为 ctx-dev、ctx-prod 和 kubernetes-admin@kubernetes，其中 current-context 表示当前的 Context 为 kubernetes-admin @kubernetes。

每个 Context 都相当于一个独立的空间。用户可以使用以下命令切换 Context：

```
[root@localhost ~]# kubectl config use-context ctx-dev
Switched to context "ctx-dev".
```

上面的命令将当前的 Context 设置为 ctx-dev，然后再次查看 Context：

```
[root@localhost ~]# kubectl config view
apiVersion: v1
clusters:
- cluster:
    certificate-authority-data: DATA+OMITTED
    server: https://192.168.2.119:6443
  name: kubernetes
contexts:
- context:
    cluster: kubernetes
    namespace: development
    user: kubernetes-admin
  name: ctx-dev
- context:
    cluster: kubernetes
    namespace: production
    user: kubernetes-admin
  name: ctx-prod
- context:
    cluster: kubernetes
    user: kubernetes-admin
  name: kubernetes-admin@kubernetes
current-context: ctx-dev
kind: Config
preferences: {}
users:
- name: kubernetes-admin
```

```
  user:
    client-certificate-data: REDACTED
    client-key-data: REDACTED
```

从上面的输出结果可知，当前的 Context 已经变成了 ctx-dev。

为了演示通过命名空间和 Context 实现资源对象的隔离，下面在当前的 Context 中创建一个 Deployment，其配置文件如下：

```
01 apiVersion: apps/v1
02 kind: Deployment
03 metadata:
04   name: nginx-deployment
05   labels:
06     app: nginx
07 spec:
08   replicas: 3
09   selector:
10     matchLabels:
11       app: nginx
12   template:
13     metadata:
14       labels:
15         app: nginx
16     spec:
17       containers:
18       - name: nginx
19         image: nginx:1.7.9
20         ports:
21         - containerPort: 80
```

将以上代码保存为 nginx-deployment.yaml，然后通过以下命令创建 Deployment：

```
[root@localhost ~]# kubectl apply -f nginx-deployment.yaml
deployment.apps/nginx-deployment created
```

接下来使用以下命令查看当前 Context 中的 Deployment 清单：

```
[root@localhost ~]# kubectl get deployments
NAME               READY   UP-TO-DATE   AVAILABLE   AGE
nginx-deployment   0/3     3            0           2m46s
```

可以发现刚刚创建的 Deployment 出现在清单当中。

接下来，使用以下命令切换到名为 ctx-prod 的 Context 中，并查看 Deployment 清单，如下所示：

```
[root@localhost ~]# kubectl config use-context ctx-prod
Switched to context "ctx-prod".
```

```
[root@localhost ~]# kubectl get deployments
No resources found in production namespace.
```

可以发现，名为 production 的命名空间中并没有前面创建的 Deployment。这说明通过命名空间和 Context，已经实现了资源对象的隔离。

10.3.2 删除命名空间

命名空间本身也是一种资源。我们可以新建命名空间，而对于不再使用的命名空间，我们需要清理掉。例如，下面的命令将名为 production 的命名空间从当前的集群中删除：

```
[root@localhost ~]# kubectl delete namespace production
namespace "production" deleted
[root@localhost ~]# kubectl get namespaces
NAME              STATUS          AGE
default           Active          12h
development       Active          4h12m
kube-node-lease   Active          12h
kube-public       Active          12h
kube-system       Active          12h
```

10.4 管理 Kubernetes 资源

在 Kubernetes 中，资源包括多种类型，例如计算资源（CPU、内存和 GPU）、存储资源（磁盘和固态硬盘）和网络资源（网络带宽、IP 地址以及端口）等。通常情况下，存储资源和网络资源管理起来相对比较容易。但是计算资源的管理，却相对比较复杂。本节将详细介绍如何对 Kubernetes 中的计算资源进行有效管理，以保证 Kubernetes 集群的稳定、高效地运行。

10.4.1 通过 requests 和 limits 属性限制资源使用

在 Kubernetes 集群中，节点是计算资源的提供者，是对资源的抽象。Pod 是计算资源的使用者，是对容器的封装。计算资源通常包括 CPU 和内存。对于 CPU 而言，它的基本计算单位为核数；对于内存而言，它的基本计算单位为字节。

当用户创建容器时，可以使用 requests 和 limits 这两个属性来分别指定容器所期望得到的 CPU 和内存的资源量，以及容器所能使用的 CPU 和内存的资源量的上限。下面分别对这两个属性进行详细介绍。

requests 是请求需要使用的计算资源数量，通常情况下，这个资源量也是保证容器能够正常运行的最低资源数量。Kubernetes 会保证容器能够使用到 requests 属性指定的资源数量。因此，请求的资源是 Kubernetes 进行调度的依据，只有当某个节点上的可用资源大于容器请求的各项资源时，调度器才会把该容器所属的 Pod 调度到该节点上。

注　意

调度器只关心节点上的可分配资源以及容器所请求的资源，而不关心节点上资源的实际使用情况。换句话说，如果节点上的容器申请的资源已经把节点上的资源用满，即使它们的使用率非常低，比如说 CPU 和内存使用率都低于 10%，调度器也不会继续调度 Pod 到该节点上。

limits 是 Pod 能够使用的资源的上限。通常情况下，requests 属性的值应该不大于 limits 属性指定的值。

requests 和 limits 属性都是可选的。如果只设置了 requests 属性，则 limits 属性的默认值被设置为当前节点的资源的最大值；如果只设置了 limits 属性，则 requests 属性的默认值被设置为与 limits 相等的值；如果 requests 和 limits 都没有设置，则在创建 Pod 时，Kubernetes 会自动使用集群的默认值。

前面提到过，CPU 的基本计算单位是核数，一个核相当于物理服务器的一个超线程内核。也就是通过以下命令查看到的 CPU 的核数：

```
[root@localhost ~]# cat /proc/cpuinfo
processor          : 0
vendor_id          : GenuineIntel
cpu family         : 6
model              : 142
model name         : Intel(R) Core(TM) i7-10510U CPU @ 1.80GHz
stepping           : 12
cpu MHz            : 2303.998
cache size         : 8192 KB
physical id        : 0
siblings           : 4
core id            : 0
cpu cores          : 4
apicid             : 0
initial apicid     : 0
fpu                : yes
fpu_exception      : yes
cpuid level        : 22
wp                 : yes
flags              : fpu vme de pse tsc msr pae mce cx8 apic sep mtrr pge mca
cmov pat pse36 clflush mmx fxsr sse sse2 ht syscall nx rdtscp lm constant_tsc
rep_good nopl xtopology nonstop_tsc cpuid tsc_known_freq pni pclmulqdq ssse3 cx16
pcid sse4_1 sse4_2 x2apic movbe popcnt aes xsave avx rdrand hypervisor lahf_lm
abm 3dnowprefetch invpcid_single fsgsbase avx2 invpcid rdseed clflushopt md_clear
flush_l1d arch_capabilities
bugs               : spectre_v1 spectre_v2 spec_store_bypass swapgs
itlb_multihit
```

```
bogomips            : 4607.99
clflush size        : 64
cache_alignment     : 64
address sizes       : 39 bits physical, 48 bits virtual
power management    :

processor           : 1
vendor_id           : GenuineIntel
cpu family          : 6
model               : 142
model name          : Intel(R) Core(TM) i7-10510U CPU @ 1.80GHz
stepping            : 12
cpu MHz             : 2303.998
cache size          : 8192 KB
physical id         : 0
siblings            : 4
core id             : 1
cpu cores           : 4
…
```

例如，用户可以使用整数值 1、2 以及 3 等分别表示 1 个 CPU 内核、2 个 CPU 内核以及 3 个 CPU 内核。除此之外，由于 Kubernetes 和 Docker 对计算资源进行了池化和虚拟化，使得用户可以指定非整数个 CPU 内核。在这种情况下，用户可以使用两种表示方式。首先是可以单独使用一个小数值。例如，用户可以在创建 Pod 时指定某个容器的 CPU 核数为 0.5，表示使用 0.5 个 CPU 内核的计算量。此外，用户还可以使用 m 作为单位，其中 1 个 CPU 内核的计算量等于 1000m。例如 0.5 个 CPU 内核可以表示为 500m。

例如，下面的代码为一个 Pod 的 YAML 配置文件：

```
01 apiVersion: v1
02 kind: Pod
03 metadata:
04   name: hello-app
05 spec:
06   containers:
07   - name: wp
08     image: wordpress
09     resources:
10       requests:
11         memory: "64Mi"
12         cpu: "250m"
13       limits:
14         memory: "128Mi"
15         cpu: "500m"
```

从第 9 行开始为资源的配置，其中第 10~12 行为 requests 属性，第 11 行指定请求的内存资源为 64MB，第 12 行指定请求的 CPU 资源为 250m，即 1/4 个 CPU 内核的计算量。第 13~15 行为 limits 属性，其中第 14 行指定内存的上限为 128MB，第 15 行指定 CPU 资源的上限为 500m，即 0.5 个 CPU 内核的计算量。

从上面的介绍可以得知，使用 requests 和 limits 这两个属性，用户为每个 Pod 分配合理的资源数量，从而能够提升整体的资源使用率。但是这个体系有个非常重要的问题需要考虑，那就是怎样准确地评估 Pod 所需要使用的资源量？如果评估得过低，会导致应用不稳定；如果过高，则会导致使用率降低。这个问题需要开发者和系统管理员共同讨论和设置。

10.4.2　通过 LimitRange 限制资源使用

在前面的内容中，我们详细介绍了使用 requests 和 limits 这 2 个属性来对 Pod 使用的计算资源进行限定。尽管这 2 个属性使用起来非常方便，但是在维护大量 Pod 的时候就显得工作量巨大。为了减轻开发人员的工作量，Kubernetes 提供了一种名称为 LimitRange 资源类型。通过 LimitRange，可以让用户为某个命名空间配置一个默认的 requests 和 limits 值，这样，在绝大部分情况下用户创建 Pod 时，就可以不用单独指定 requests 和 limits 值了。

为了能够使得读者快速地理解 LimitRange 的创建方法，下面首先看一个简单的例子。

```
01 apiVersion: "v1"
02 kind: "LimitRange"
03 metadata:
04   name: devel-limits
05 spec:
06   limits:
07     - type: "Container"
08       max:
09         cpu: "2"
10         memory: "1Gi"
11       min:
12         cpu: "100m"
13         memory: "4Mi"
14       default:
15         cpu: "500m"
16         memory: "200Mi"
17       defaultRequest:
18         cpu: "200m"
19         memory: "100Mi"
```

第 2 行指定所创建的资源的类型为 LimitRange。第 4 行指定 LimitRange 的名称为 devel-limits。从第 5 行开始为 LimitRange 的规格描述，其中最重要的是 limits。第 7 行指定当前 LimitRange 所应用的资源类型为 Container，即容器。第 8 行的 max 属性指定了命名空间中所有容器的 limits 属性值的上限。第 11 行的 min 属性指定了命名空间中所有容器的 requests

属性值的下限。第 14 行 default 为默认的 limits 属性值，即如果在创建容器时没有单独指定 limits 属性值，则采用 CPU 资源量为 500m，内存量为 200Mi。第 17 行的 defaultRequest 属性指定了所有容器默认的 requests 的资源量。

将以上文件保存为 devel-limits.yaml，然后使用以下命令创建 LimitRange 资源对象：

```
[root@localhost ~]# kubectl apply -f devel-limits.yaml
--namespace=development
limitrange/devel-limits created
```

其中--namespace 选项用来指定 LimitRange 所在命名空间。在本例中，development 命名空间已经在前面创建。

创建完成之后，使用以下命令查看 development 命名空间中的 LimitRange 列表：

```
[root@localhost ~]# kubectl get limitranges --namespace=development
NAME            CREATED AT
devel-limits    2020-12-05T08:44:22Z
```

从上面的代码可以得知，通过 LimitRange，可以非常方便地定义容器的计算资源的各项指标。下面对 LimitRange 的配置方法进行详细介绍。

在 LimitRange 中，所面向的资源不仅仅是 Container，还可以是 Pod。因此，type 属性可以取 Container 和 Pod 这 2 个值。

在 LimitRange 的规格描述中，用户可以使用 min、max 和 maxLimitRequestRatio 这 3 个属性。对于 Container 而言，用户还可以使用 defaultRequest 和 defaultLimit。下面分别介绍这些属性的含义。

对于 Container 来说，min 属性是指特定命名空间中所有容器的 requests 属性值的下限，max 属性是指特定命名空间中所有容器的 limits 属性值的上限。maxLimitRequestRatio 是一个比值，它是 limits 值和 requests 值的比值。由于资源调度都是基于 requests 的值，因此可能会出现资源超售情况，这个比值显示了超售的比例。defaultRequest 属性是命名空间中所有的容器的默认的 requests 的值，如果在创建容器时没有指定 requests 属性的资源量，则使用 defaultRequest 的值。defaultLimit 是命名空间中所有容器的默认的 limits 的值，如果在创建容器时没有指定 limits 的值，则使用默认值。

对于 Pod 而言，min 属性是指 Pod 中所有容器的 requests 属性值总和的下限，max 属性是指 Pod 中所有容器的 limits 属性值总和的上限。maxLimitRequestRatio 属性则限制了 Pod 中所有容器的 limits 属性值总和与 requests 属性值总和的比例的上限。

下面通过创建一个 Pod 来说明如何触发 LimitRange 的限制，该 Pod 的 YAML 配置文件如下：

```
01 apiVersion: v1
02 kind: Pod
03 metadata:
04   name: sakura
05   labels:
06     app: nginx
07 spec:
```

```
08   containers:
09   - image: nginx
10     imagePullPolicy: IfNotPresent
11     name: hawk
12     resources:
13       limits:
14         cpu: 3
15         memory: 250Mi
```

在前面的 LimitRange 中，指定容器的 CPU 核数最大为 2。为了触发该 LimitRange，第 14 行将新建的容器的 CPU 核数指定为 3。将以上代码保存为 invalid-pod.yaml，然后使用以下命令尝试创建该 Pod：

```
[root@localhost ~]# kubectl apply -f invalid-pod.yaml
Error from server (Forbidden): error when creating "invalid-pod.yaml": pods "sakura" is forbidden: maximum cpu usage per Container is 2, but limit is 3
```

很明显，上面的命令执行失败。由于所请求的 CPU 资源超出了 LimitRange 的限制，因此 Kubernetes 禁止用户创建该 Pod，这也说明前面创建的名为 devel-limits 的 LimitRange 发挥了作用。

10.4.3　资源配额

前面介绍的方法可以对命名空间中的容器进行资源利用方面的限制。资源配额可以对整个命名空间所能够使用的资源总额度进行限制。这里的资源不仅仅包括 CPU、内存等计算资源，还可以包括 Kubernetes 中的资源对象，例如 Pod 和服务的数量等。通过资源配额，Kubernetes 可以防止某个命名空间下的用户，不加限制地使用超过期望的资源，比如说不对资源进行评估就大量申请高性能的 Pod。

下面是一个资源配额的实例，它限制了当前命名空间只能使用少于或者等于 20 个 CPU 内核以及不超过 1GB 的内存，并且最多只能创建 10 个 Pod、20 个 RC、5 个服务：

```
01 apiVersion: v1
02 kind: ResourceQuota
03 metadata:
04   name: devel-quota
05 spec:
06   hard:
07     cpu: 20
08     memory: 1Gi
09     pods: 10
10     replicationcontrollers: 20
11     resourcequotas: 1
12     services: 5
```

将以上代码保存为 devel-quota.yaml，然后使用以下命令创建：

```
[root@localhost ~]# kubectl apply -f devel-quota.yaml --namespace=development
resourcequota/quota created
[root@localhost ~]# kubectl get resourcequota --namespace=development
NAME    AGE     REQUEST                                                LIMIT
quota   3h58m   cpu: 450m/20, memory: 164Mi/1Gi, pods: 5/10,
replicationcontrollers: 0/20, resourcequotas: 1/1, services: 0/5
```

从上面的输出结果可知，当前命名空间中已经使用了450m的CPU资源，一共可以使用20个CPU内核；内存一共使用了164MB，可以使用1GB；最多可以创建10个Pod，目前已经创建了5个。同样replicationcontrollers和services等也都有限制。

资源配额能够配置的选项还很多，例如GPU、存储、Configmaps、PersistentVolumeClaims等等，更多信息可以参考官方文档。

资源配额要解决的问题和使用方法都相对独立和简单，但是它也有一个限制，那就是它不能根据集群资源动态伸缩。一旦配置之后，资源配额就不会改变，即使集群增加了节点，整体资源增多也没有用。Kubernetes现在没有解决这个问题，但是用户可以通过编写一个Controller的方式来自己实现。

10.4.4 资源服务质量管理

requests和limits的配置除了表明资源情况和限制资源使用之外，还有一个隐藏的作用，即它决定了Pod的资源服务质量等级。

前面已经介绍过，如果在创建Pod时没有指定limits属性，则该Pod可以使用所在节点上任意多的可用资源。这类Pod能够灵活地使用资源，但这也导致它不稳定且危险。对于这类Pod，用户一定要在它占用过多资源而导致节点资源紧张时将其处理掉。优先处理这类Pod，而不是处理资源使用处于自己请求范围内的Pod。Pod的服务质量控制的含义就是根据Pod的资源请求，把Pod分成不同的重要等级。

Kubernetes把Pod分成了三个服务质量等级：

- Guaranteed：优先级最高，除非Pod使用超过了它们的limits属性设置的数值，或者节点的内存压力很大而且没有级别更低的Pod，否则这类Pod就不会被"杀死"。用户可以在该级别的Pod上运行数据库应用或者一些重要的业务应用。
- Burstable：这种类型的Pod可以使用多于自己请求的资源，上限由limits指定，如果limits没有配置，则可以使用主机的任意可用资源。但是处于该级别的Pod重要性认为比较低，可以是一般性的应用或者批处理任务。
- Best Effort：优先级最低，集群不知道Pod的资源请求情况，调度不考虑资源，可以运行到任意节点上，可以是一些临时性的不重要的应用。Pod可以使用节点上任何可用资源，但在资源不足时也会被优先"杀死"。

通常情况下，如果在创建资源时，没有设置requests和limits属性，则资源的服务质量优先级最低，在节点资源紧张的情况下，最容易被优先"杀死"。

10.5　Pod 驱逐机制

前面所介绍的都是理想情况下的 Kubernetes 工作状况，在理想情况下，Kubernetes 集群中的各种资源都是足够用的，并且所有的应用系统都是在使用规定范围内的资源。但是在生产环境中，用户所面临的情况却不会如此简单，而是会经常遇到资源不足的情况。此时，用户需要保证整个集群的可用，并且尽可能减少损失。Kubernetes 提供了一种 Pod 驱逐机制，来满足这种需求。本节将详细介绍 Pod 驱逐的原理及使用方法。

10.5.1　驱逐触发条件

现实环境中，Pod 驱逐机制用于在管理集群的时候常常会遇到资源不足的情况。在这种情况下，用户最大的任务首先就是要保证整个集群处于可用状态，然后就是尽可能减少应用的损失。保证集群可用比较容易理解，首先要保证系统层面的核心进程正常，其次要保证 Kubernetes 本身组件进程不出问题。至于减少应用的损失，最常用的方法就是尽量“杀死”不重要的应用，让重要的应用不受影响，即 Pod 驱逐。当然，这涉及 Pod 的优先级问题。

Pod 的驱逐是在 kubelet 中实现的，因为 kubelet 能动态地感知到节点上资源使用率的实时变化情况。一旦发现某个不可压缩资源出现要耗尽的情况，就会主动终止节点上的 Pod，让节点能够正常运行。被终止的 Pod 中的所有容器会停止，状态会被设置为 Failed。

那么，哪些资源不足会导致 kubelet 执行 Pod 驱逐呢？目前主要有三种情况，分别是实际内存不足、节点文件系统的可用空间不足，以及镜像文件系统的可用空间不足。在 Kubernetes 中，这些称为驱逐信号，分别使用 memory.avaliable、nodefs.avaliable、nodefs.inodesFree、imagefs.avaliable 以及 imagefs.inodesFree 表示。

有了数据的来源，另外一个问题是 Pod 驱逐触发的时机，也就是到什么程度需要触发 Pod 驱逐。Kubernetes 支持两种模式，分别为按照百分比和按照绝对数量。例如，对于一个拥有 32G 内存的节点，当可用内存少于 10%时启动驱逐程序，可以配置 memory.available<10%或者 memory.available<3.2Gi。前者为百分比表示，后者为绝对数量表示。

注　意

默认情况下，kubelet 的驱逐规则是 memory.available<100Mi，对于生产环境来说，这个配置是不可接受的，所以一定要根据实际情况进行修改。

10.5.2　软驱逐和硬驱逐

通过前面的介绍可以得知，尽管 Pod 驱逐保证了重要应用的正常运行，但是对于被标记为不重要的应用来说，这是一种具有毁灭性的行为，因此，在使用的时候必须谨慎。有时候内存使用率增高只是暂时性的，有可能 20s 就能恢复，这时候启动 Pod 驱逐程序意义不大，而且可能会导致应用的不稳定，用户需要考虑到这种情况应该如何处理。有的时候内存使用率过高，比如高于 95%，那么我们不应该再多作评估和考虑，而是赶紧启动 Pod 驱逐程序，因为这种情况再花费时间去判断，可能会导致内存继续增长，系统完全崩溃。

为了解决这个问题，Kubernetes 引入了软驱逐和硬驱逐的概念。

软驱逐可以在资源紧缺，但没有发展到非常严重的时候触发，例如内存使用率为 85%。软驱逐在触发以后不会立即驱逐 Pod，而是继续观察一段时间，如果资源使用率高于阈值的情况持续一定时间，才开始驱逐。并且驱逐 Pod 的时候，也不会立即“杀死”Pod，而是预留一定的时间让 Pod 完成清理工作，自己停止。如果超过指定的时间，Pod 还没有自动终止，才会主动“杀死”Pod。

和软驱逐相关的启动参数是：

- --eviction-soft：软驱逐触发条件，比如 memory.available<1Gi。
- --eviction-soft-grace-period：触发条件持续多久才开始驱逐，比如 memory.available=2m30s。
- --eviction-max-Pod-grace-period：正式“杀死”Pod 前等待 Pod 自动停止的时间，如果到时间还没有结束，就直接将其“杀死”。

前面两个参数必须同时配置，软驱逐才能正常工作。后一个参数会和 Pod 本身配置的等待时间比较，选择较小的一个生效。

硬驱逐则比较简单，当 kubelet 发现节点达到配置的硬驱逐阈值后，立即开始驱逐程序，并且不会等待 Pod 完成清理工作，也就是说立即强制“杀死”Pod。因此，硬驱逐仅仅用在非常紧急的情况下。其对应的配置参数只有一个--evictio-hard。

设置这两种驱逐机制是为了平衡节点稳定性和对 Pod 的影响，软驱逐照顾到了 Pod 的优雅退出，减少驱逐对 Pod 的影响；而硬驱逐则照顾到节点的稳定性，防止资源的快速消耗导致节点不可用。

软驱逐和硬驱逐可以单独配置，不过还是推荐两者都进行配置，一起使用。

10.5.3 驱逐优先级

前面已经介绍过，Pod 驱逐的重要原则是尽量减少对应用程序的影响。如果是存储资源不足，kubelet 会根据情况清理状态为 Dead 的 Pod 和它的所有容器，以及清理所有没有使用的镜像。如果上述清理并没有让节点恢复正常，那么 kubelet 就开始清理 Pod。

通常情况下，系统组件的 Pod 要比普通的 Pod 更重要，另外运行数据库的 Pod 自然要比运行一个无状态应用的 Pod 更重要。kubelet 会根据 Pod 的 requests 和 limits、优先级、以及 Pod 实际的资源使用情况来判断 Pod 的驱逐优先级。

简单来说，kubelet 会根据以下标准对 Pod 进行排序：Pod 是否使用了超过请求的紧张资源、Pod 的优先级，以及使用的紧缺资源和请求的紧张资源之间的比例。

10.5.4 防止波动

Kubernetes 的 Pod 驱逐波动有两种情况，下面分别进行介绍。

首先第一种情况是在驱逐条件发出之后，如果 kubelet 驱逐一部分 Pod，让资源使用率低于阈值就停止，那么很可能过一段时间资源使用率又会达到阈值，从而再次发出驱逐，如此循环往复。对于这个问题，用户可以使用--eviction-minimum-reclaim 选项来解决，这个参数配置每次驱逐至少要清理出来多少资源才会停止。通过这个选项，可以在驱逐时为节点的各种资源预留一定的增长空间，避免频繁地触发 Pod 驱逐。

另外一种波动情况就是 Pod 被驱逐之后并不会从此消失不见，常见的情况是 Kubernetes 会自动生成一个新的 Pod 来取代，并经过调度选择一个节点继续运行。如果不做额外地处理，Kubernetes 将新的 Pod 调度到原来节点的可能性比较大。如果被驱逐的 Pod 再次调度到原来的节点，很可能会再次触发驱逐程序，然后 Pod 再次被调度到当前节点，循环往复。对于这种情况，Kubernetes 的解决方法是发生 Pod 驱逐之后，kubelet 立即更新节点的状态，调度器感知到这一情况，暂时不往该节点调度 Pod 即可。用户可以使用--eviction-pressure-transition-period 参数来指定 kubelet 多久才上报节点的状态，因为默认的上报状态周期比较短，频繁更改节点状态会导致驱逐波动。

10.6　Kubernetes 集群的高可用部署方案

在生产环境中， 管理员必须保证 Kubernetes 集群运行的稳定性，避免由于 Kubernetes 集群故障而导致业务方面的巨大损失。Kubernetes 集群的高可用性有多种实现方法，其中使用 kubeadmin 工具部署是最简洁的一种方式。本节将详细介绍如何使用 kubeadmin 工具实现 Kubernetes 集群的高可用性。

10.6.1　Kubernetes 集群的高可用性原理

在 Kubernetes 集群中主要有两种节点，分别为 Master 节点和工作节点（Node）。其中 Master 节点为管理节点，管理整个 Kubernetes 集群，接收外部命令，维护集群状态。如果 Master 节点出现故障，则整个集群会失去控制。在 Master 节点上面，运行的主要服务有 apiserver、etcd、scheduler 以及 controller-manager。工作节点（Node 节点），主要执行计算任务，运行的主要服务有 kubelet 和 kube-proxy。当工作节点出现故障，Kubernetes 会将 Pod 调度到其他的工作节点上，并不会影响整个集群的运行，甚至对应用系统的影响也非常小。因此，Kubernetes 集群的高可用主要是指 Master 节点的高可用。

在 Master 节点中，apiserver 的功能是作为 API 服务器，所有外部与 Kubernetes 集群的交互都需要经过它。apiserver 可以同时存在多个，并且通过负载均衡器实现高可用性。etcd 本身是一个高可用的分布式“键-值”存储系统，可以实现集群。scheduler 的功能是将 Pod 调度到具体的工作节点上，而 controller-manager 的功能是执行控制器逻辑，通过 apiserver 监控集群状态，做出相应的处理。scheduler 和 controller-manager 这两个服务在一个 Kubernetes 集群中只会有一个处于激活状态。如果存在着多个 Master 节点，则它们会依据相应的算法选举产生处于激活状态的节点。

etcd 的部署方式主要有两种，其中一种为堆叠式的 etcd 集群，在这种方式中，etcd 分布式数据存储集群堆叠在 Master 节点中。另外一种为外部式，此时，etcd 独立于 Master 节点。图 10-1 所示显示了堆叠式的 etcd 集群方案。

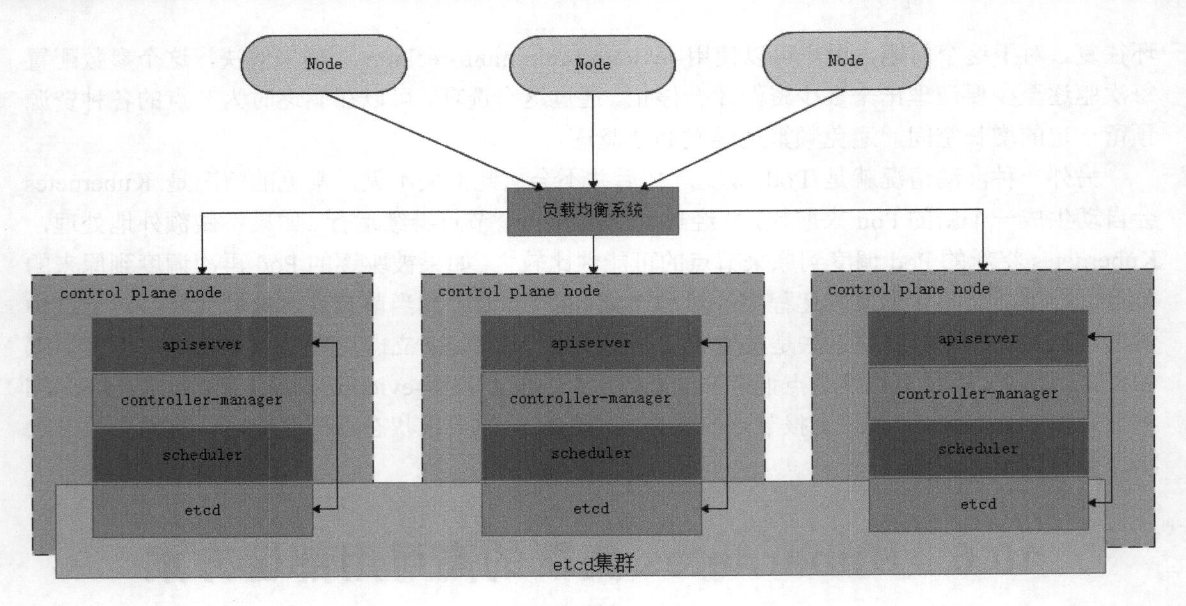

图 10-1　堆叠式 etcd 集群

10.6.2　安装环境准备

在生产环境中，无论是 Master 节点，还是工作节点（Node 节点），通常都是物理服务器。在本例中，为了便于演示，使用 VirtualBox 虚拟机软件创建了 4 个虚拟机。这 4 个虚拟机的内存都是 2GB，虚拟 CPU 为 2 个，网络连接方式为桥接网卡，安装的操作系统都是 CentOS 8。表 10-1 列出了这 4 台虚拟机 IP 地址的分配情况。

表 10-1　虚拟机IP地址分配

主　机　名	IP 地址	虚拟 IP 地址
master1	192.168.2.128	192.168.1.132
master2	192.168.2.129	
master3	192.168.2.130	
node1	192.168.2.131	

为了能够顺利部署 Kubernetes，用户还需要做一些配置工作，下面分步进行介绍，以下步骤需要在所有的虚拟机上面完成操作。

（1）配置 CentOS 软件源。为了提高安装速度，用户需要将 CentOS 默认的软件源，修改为阿里镜像的软件源。

```
[root@master1 ~]# rm -rfv /etc/yum.repos.d/*
removed '/etc/yum.repos.d/CentOS-AppStream.repo'
removed '/etc/yum.repos.d/CentOS-Base.repo'
removed '/etc/yum.repos.d/CentOS-centosplus.repo'
removed '/etc/yum.repos.d/CentOS-CR.repo'
removed '/etc/yum.repos.d/CentOS-Debuginfo.repo'
```

```
    removed '/etc/yum.repos.d/CentOS-Devel.repo'
    removed '/etc/yum.repos.d/CentOS-Extras.repo'
    removed '/etc/yum.repos.d/CentOS-fasttrack.repo'
    removed '/etc/yum.repos.d/CentOS-HA.repo'
    removed '/etc/yum.repos.d/CentOS-Media.repo'
    removed '/etc/yum.repos.d/CentOS-PowerTools.repo'
    removed '/etc/yum.repos.d/CentOS-Sources.repo'
    removed '/etc/yum.repos.d/CentOS-Vault.repo'
    [root@master1 ~]# curl -o /etc/yum.repos.d/CentOS-Base.repo
http://mirrors.aliyun.com/repo/Centos-8.repo
```

curl 命令的功能是将远程的软件源配置文件下载到本地的指定目录。

（2）配置主机名。为了能够正确解析主机名，用户需要在/etc/hosts 文件中进行相应的配置，增加以下代码：

```
192.168.2.128       master1
192.168.2.129       master2
192.168.2.130       master3
192.168.2.131       node1
```

（3）关闭交换分区，并且在/etc/fstab 文件中删除交换分区的配置。

首先执行以下命令关闭交换分区：

```
[root@master1 ~]# swapoff -a
```

然后打开/etc/fstab 文件，找到以下代码，将该行的前面添加注释符号：

```
/dev/mapper/cl-swap     swap                    swap    defaults        0 0
```

（4）配置内核参数，将桥接的 IPv4 流量传递到 iptables 的链，命令如下：

```
[root@master1 ~]# cat > /etc/sysctl.d/k8s.conf <<EOF
net.bridge.bridge-nf-call-ip6tables = 1
net.bridge.bridge-nf-call-iptables = 1
EOF
```

然后执行以下命令使得上面的配置生效：

```
    [root@node1 ~]# sysctl --system
    * Applying /usr/lib/sysctl.d/10-default-yama-scope.conf ...
    kernel.yama.ptrace_scope = 0
    * Applying /usr/lib/sysctl.d/50-coredump.conf ...
    kernel.core_pattern =
|/usr/lib/systemd/systemd-coredump %P %u %g %s %t %c %h %e
    * Applying /usr/lib/sysctl.d/50-default.conf ...
    kernel.sysrq = 16
    kernel.core_uses_pid = 1
    kernel.kptr_restrict = 1
```

```
net.ipv4.conf.all.rp_filter = 1
net.ipv4.conf.all.accept_source_route = 0
net.ipv4.conf.all.promote_secondaries = 1
net.core.default_qdisc = fq_codel
fs.protected_hardlinks = 1
fs.protected_symlinks = 1
* Applying /usr/lib/sysctl.d/50-libkcapi-optmem_max.conf ...
net.core.optmem_max = 81920
* Applying /usr/lib/sysctl.d/50-pid-max.conf ...
kernel.pid_max = 4194304
* Applying /etc/sysctl.d/99-sysctl.conf ...
* Applying /etc/sysctl.d/k8s.conf ...
* Applying /etc/sysctl.conf ...
```

（5）禁用 SELinux。

首先执行以下命令禁用 SELinux：

```
[root@master1 ~]# setenforce 0
```

然后修改/etc/selinux/config 文件，将其中的 SELINUX 选项的值设置为 disabled，代码如下：

```
SELINUX=disabled
```

（6）关闭防火墙，命令如下：

```
[root@master1 ~]# systemctl stop firewalld
[root@master1 ~]# systemctl disable firewalld
```

（7）安装所需要的软件包，命令如下：

```
[root@master1 ~]# yum install -y yum-utils device-mapper-persistent-data lvm2
```

10.6.3 安装 Master 节点

接下来，再详细介绍 master1 节点的安装步骤。用户需要在 master1、master2 和 master3 这 3 个 Master 节点上面执行以下步骤。

（1）添加 Docker CE 阿里镜像软件源，命令如下：

```
[root@master1 ~]# yum-config-manager --add-repo
https://mirrors.aliyun.com/docker-ce/linux/centos/docker-ce.repo
Adding repo from:
https://mirrors.aliyun.com/docker-ce/linux/centos/docker-ce.repo
```

（2）安装 Docker CE，命令如下：

```
[root@master1 ~]# yum -y install docker-ce
```

如果在安装的时候出现以下错误，请升级 containerd.io 为 1.44.1 以上。

```
    Problem: package docker-ce-3:20.10.0-3.el7.x86_64 requires containerd.io >=
1.4.1, but none of the providers can be installed
```

成功安装 Docker CE 之后，启用并启动该服务，命令如下：

```
    [root@master1 ~]# systemctl start docker
    [root@master1 ~]# systemctl enable docker
    Created symlink /etc/systemd/system/multi-user.target.wants/docker.service
→ /usr/lib/systemd/system/docker.service.
```

（3）安装 kubectl、kubelet、kubeadm。

为了加快安装速度，需要添加阿里的 Kubernetes 镜像软件源，命令如下：

```
    [root@master1 ~]# cat <<EOF > /etc/yum.repos.d/kubernetes.repo
    [kubernetes]
    name=Kubernetes
    baseurl=https://mirrors.aliyun.com/kubernetes/yum/repos/kubernetes-el7-x8
6_64/
    enabled=1
    gpgcheck=1
    repo_gpgcheck=1
    gpgkey=https://mirrors.aliyun.com/kubernetes/yum/doc/yum-key.gpg
https://mirrors.aliyun.com/kubernetes/yum/doc/rpm-package-key.gpg
    EOF
```

然后安装 Kubernetes 相关组件，命令如下：

```
    [root@master1 ~]# yum install kubectl kubelet kubeadm -y
```

启用 kubelet，命令如下：

```
    [root@master1 ~]# systemctl enable kubelet
    Created symlink
/etc/systemd/system/multi-user.target.wants/kubelet.service →
/usr/lib/systemd/system/kubelet.service
```

10.6.4　安装 haproxy

haproxy 是一个高性能的负载均衡系统。在本例中，通过 haproxy 实现 3 个 apiserver 的负载均衡。以下操作分别在 master1、master2 和 master3 上进行。

（1）安装 haproxy，命令如下：

```
    [root@master1 ~]# yum -y install haproxy
```

（2）修改配置文件。修改 haproxy 的配置文件/etc/haproxy/haproxy.cfg，代码如下：

```
    01  global
    02      log         /var/log/haproxy.log local0
```

```
03      pidfile     /var/run/haproxy.pid
04      maxconn     4000
05      user        haproxy
06      group       haproxy
07      daemon
08  defaults
09     mode                        http
10     log                         global
11     retries                     1
12     timeout http-request        10s
13     timeout queue               20s
14     timeout connect             5s
15     timeout client              20s
16     timeout server              20s
17     timeout http-keep-alive 10s
18     timeout check               10s
19  listen admin_stats
20     mode                        http
21     bind                        0.0.0.0:1080
22     log                         127.0.0.1 local0 err
23     stats refresh               30s
24     stats uri                   /haproxy-status
25     stats realm                 Haproxy\ Statistics
26     stats auth                  admin:admin
27     stats hide-version
28     stats admin if TRUE
29
30  frontend apiserver
31    bind *:8443
32    mode tcp
33    option tcplog
34    default_backend apiserver
35
36  backend apiserver
37     option httpchk GET /healthz
38     http-check expect status 200
39     mode tcp
40     option ssl-hello-chk
41     balance    roundrobin
42     server master1  192.168.1.3:6443 weight 1 maxconn 1000 check inter 2000
rise 2 fall 3
43     server master2  192.168.1.4:6443 weight 1 maxconn 1000 check inter 2000
rise 2 fall 3
```

```
44      server master3  192.168.1.5:6443 weight 1 maxconn 1000 check inter 2000
rise 2 fall 3
```

第 26 行指定了 haproxy 统计平台的账号和密码都是 admin。第 30 行开始指定 haproxy 的前端配置选项，包括监听的端口、工作模式以及默认的后端服务器。第 36 行开始配置 haproxy 的后端选项，其中从 42 行开始分别指定前面配置的 3 个 Master 节点。

（3）启动并启用 haproxy，命令如下：

```
[root@master1 haproxy]# systemctl start haproxy
[root@master1 haproxy]# systemctl enable haproxy
```

10.6.5　安装 keepalived

keepalived 是一个非常流行的服务器高可用解决方案。在本例中，通过 keepalived 实现 apiserver 的故障转移。

keepalived 的安装方法如下：

```
[root@master1 haproxy]# yum -y install keepalived
```

修改 keepalived 的主配置文件/etc/keepalived/keepalived.conf，代码如下：

```
01 ! /etc/keepalived/keepalived.conf
02 ! Configuration File for keepalived
03 global_defs {
04     router_id LVS_K8S
05 }
06 vrrp_script check_apiserver {
07   script "/etc/keepalived/check_apiserver.sh"
08   interval 3
09   weight -2
10   fall 10
11   rise 2
12 }
13
14 vrrp_instance VI_1 {
15     state MASTER
16     interface enp0s3
17     virtual_router_id 51
18     priority 100
19     authentication {
20         auth_type PASS
21         auth_pass kubernetes
22     }
23     virtual_ipaddress {
24         192.168.1.7
```

```
25      }
26      track_script {
27          check_apiserver
28      }
29  }
```

第 7 行配置了一个名称为 check_apiserver.sh 的脚本文件，该文件的功能是定期检查 apiserver 的进程是否存在。第 16 行指定服务的网络接口为 enp0s3，用户需要根据自己的实际情况进行修改。第 23 行配置了一个虚拟 IP 地址，该虚拟 IP 地址将作为 3 个 apiserver 对外服务的 IP 地址。

接下来编写/etc/keepalived/check_apiserver.sh 脚本文件，代码如下：

```
01  #!/bin/sh
02
03  errorExit() {
04      echo "*** $*" 1>&2
05      exit 1
06  }
07
08  curl --silent --max-time 2 --insecure https://localhost:8443/ -o /dev/null
|| errorExit "Error GET https://localhost:8443/"
09  if ip addr | grep -q 192.168.1.7; then
10      curl --silent --max-time 2 --insecure https://192.168.17:8443/ -o
/dev/null || errorExit "Error GET https://192.168.1.7:8443/"
11  fi
```

其中的 192.168.1.7 即为 keepalived 的虚拟 IP 地址。授予所有用户执行 check_apiserver.sh 的权限，命令如下：

```
[root@master2 ~]# chmod +x /etc/keepalived/check_apiserver.sh
```

然后启动并启用 keepalived，命令如下：

```
[root@master2 ~]# systemctl enable keepalived
[root@master2 ~]# systemctl start keepalived
```

10.6.6 查看 haproxy 统计报告

当 haproxy 配置完成之后，用户就可以通过浏览器访问统计报告页面，网址如下：

http://192.168.1.7:1080/haproxy-status

其中 192.168.1.7 为 keepalived 的虚拟 IP 地址，服务端口 1080 即 haproxy.cfg 配置文件中配置的监听端口。用户名和密码都是 admin，这个是在 haproxy.cfg 配置文件中的第 26 行指定的，用户可以根据自己的实际情况进行修改。

haproxy 的统计报告如图 10-2 所示。

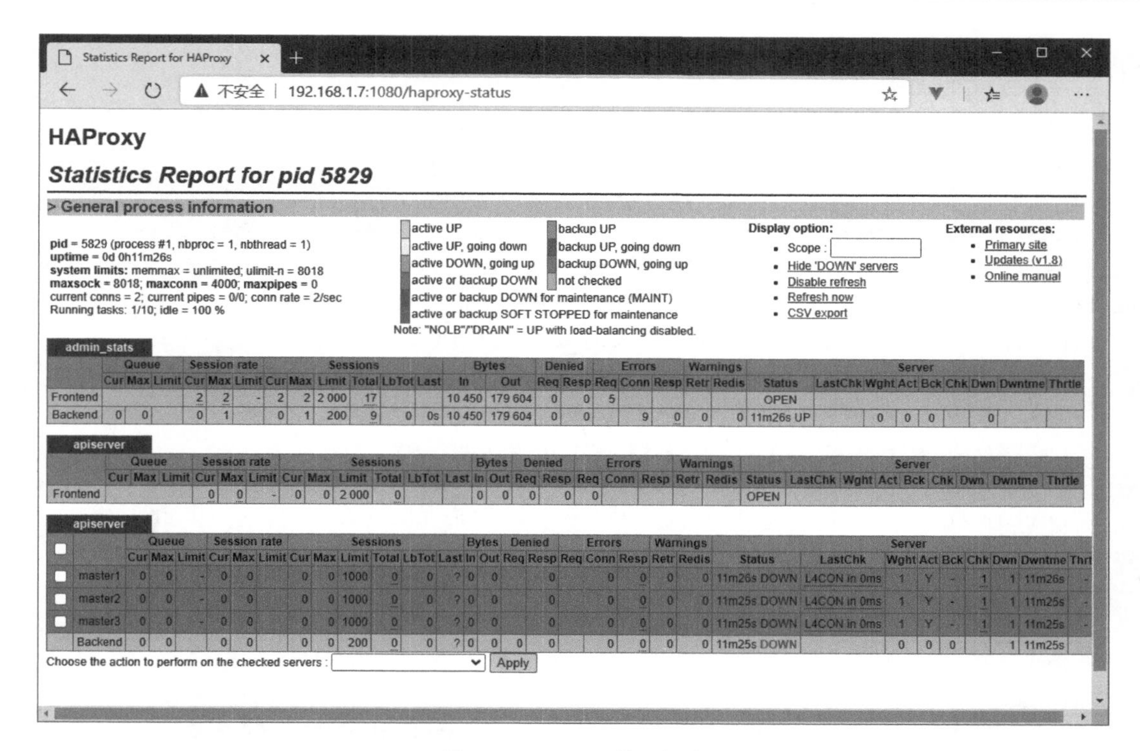

图 10-2 haproxy 统计报告

从图 10-2 可以看出，目前 3 个 Master 节点都是红色的，其状态为 DOWN，这是因为我们目前还没有初始化 Master 节点。等初始化完成之后，这 3 个 Master 节点会变成绿色。

10.6.7 初始化 Master 节点

接下来进行 Master 节点的初始化操作，需要特别注意，只需要初始化一个 Master 节点就可以了，其余的 Master 节点是通过 kubeadm join 命令加入的。

master1 的初始化命令如下：

```
    [root@master1 ~]# kubeadm init --kubernetes-version=1.20.0
--apiserver-advertise-address=192.168.1.3
--control-plane-endpoint=192.168.1.7:8443 --image-repository
registry.aliyuncs.com/google_containers  --service-cidr=10.10.0.0/16
--pod-network-cidr=10.122.0.0/16 --upload-certs
    [init] Using Kubernetes version: v1.20.0
    [preflight] Running pre-flight checks
            [WARNING IsDockerSystemdCheck]: detected "cgroupfs" as the Docker
cgroup driver. The recommended driver is "systemd". Please follow the guide at
https://kubernetes.io/docs/setup/cri/
            [WARNING FileExisting-tc]: tc not found in system path
    [preflight] Pulling images required for setting up a Kubernetes cluster
    [preflight] This might take a minute or two, depending on the speed of your
internet connection
```

```
    [preflight] You can also perform this action in beforehand using 'kubeadm config
images pull'
    [certs] Using certificateDir folder "/etc/kubernetes/pki"
    [certs] Generating "ca" certificate and key
    [certs] Generating "apiserver" certificate and key
    [certs] apiserver serving cert is signed for DNS names [kubernetes
kubernetes.default kubernetes.default.svc kubernetes.default.svc.cluster.local
master1] and IPs [10.10.0.1 192.168.1.3 192.168.1.7]
    [certs] Generating "apiserver-kubelet-client" certificate and key
    [certs] Generating "front-proxy-ca" certificate and key
    [certs] Generating "front-proxy-client" certificate and key
    [certs] Generating "etcd/ca" certificate and key
    [certs] Generating "etcd/server" certificate and key
    [certs] etcd/server serving cert is signed for DNS names [localhost master1]
and IPs [192.168.1.3 127.0.0.1 ::1]
    [certs] Generating "etcd/peer" certificate and key
    [certs] etcd/peer serving cert is signed for DNS names [localhost master1]
and IPs [192.168.1.3 127.0.0.1 ::1]
    [certs] Generating "etcd/healthcheck-client" certificate and key
    [certs] Generating "apiserver-etcd-client" certificate and key
    [certs] Generating "sa" key and public key
    [kubeconfig] Using kubeconfig folder "/etc/kubernetes"
    [endpoint] WARNING: port specified in controlPlaneEndpoint overrides bindPort
in the controlplane address
    [kubeconfig] Writing "admin.conf" kubeconfig file
    [endpoint] WARNING: port specified in controlPlaneEndpoint overrides bindPort
in the controlplane address
    [kubeconfig] Writing "kubelet.conf" kubeconfig file
    [endpoint] WARNING: port specified in controlPlaneEndpoint overrides bindPort
in the controlplane address
    [kubeconfig] Writing "controller-manager.conf" kubeconfig file
    [endpoint] WARNING: port specified in controlPlaneEndpoint overrides bindPort
in the controlplane address
    [kubeconfig] Writing "scheduler.conf" kubeconfig file
    [kubelet-start] Writing kubelet environment file with flags to file
"/var/lib/kubelet/kubeadm-flags.env"
    [kubelet-start] Writing kubelet configuration to file
"/var/lib/kubelet/config.yaml"
    [kubelet-start] Starting the kubelet
    [control-plane] Using manifest folder "/etc/kubernetes/manifests"
    [control-plane] Creating static Pod manifest for "kube-apiserver"
    [control-plane] Creating static Pod manifest for "kube-controller-manager"
    [control-plane] Creating static Pod manifest for "kube-scheduler"
```

```
    [etcd] Creating static Pod manifest for local etcd in "/etc/kubernetes/
manifests"
    [wait-control-plane] Waiting for the kubelet to boot up the control plane as
static Pods from directory "/etc/kubernetes/manifests". This can take up to 4m0s
    [apiclient] All control plane components are healthy after 28.117852 seconds
    [upload-config] Storing the configuration used in ConfigMap "kubeadm-config"
in the "kube-system" Namespace
    [kubelet] Creating a ConfigMap "kubelet-config-1.20" in namespace kube-system
with the configuration for the kubelets in the cluster
    [upload-certs] Storing the certificates in Secret "kubeadm-certs" in the
"kube-system" Namespace
    [upload-certs] Using certificate key:
    d87854ac7a180dd5c9720336f6700d76c2e7e998317f7b7e4650d6e90758beff
    [mark-control-plane] Marking the node master1 as control-plane by adding the
labels "node-role.kubernetes.io/master=''" and
"node-role.kubernetes.io/control-plane='' (deprecated)"
    [mark-control-plane] Marking the node master1 as control-plane by adding the
taints [node-role.kubernetes.io/master:NoSchedule]
    [bootstrap-token] Using token: 87z2c8.rxtz78pfculvq5il
    [bootstrap-token] Configuring bootstrap tokens, cluster-info ConfigMap, RBAC
Roles
    [bootstrap-token] configured RBAC rules to allow Node Bootstrap tokens to get
nodes
    [bootstrap-token] configured RBAC rules to allow Node Bootstrap tokens to post
CSRs in order for nodes to get long term certificate credentials
    [bootstrap-token] configured RBAC rules to allow the csrapprover controller
automatically approve CSRs from a Node Bootstrap Token
    [bootstrap-token] configured RBAC rules to allow certificate rotation for all
node client certificates in the cluster
    [bootstrap-token] Creating the "cluster-info" ConfigMap in the "kube-public"
namespace
    [kubelet-finalize] Updating "/etc/kubernetes/kubelet.conf" to point to a
rotatable kubelet client certificate and key
    [addons] Applied essential addon: CoreDNS
    [endpoint] WARNING: port specified in controlPlaneEndpoint overrides bindPort
in the controlplane address
    [addons] Applied essential addon: kube-proxy

    Your Kubernetes control-plane has initialized successfully!

    To start using your cluster, you need to run the following as a regular user:

      mkdir -p $HOME/.kube
```

```
  sudo cp -i /etc/kubernetes/admin.conf $HOME/.kube/config
  sudo chown $(id -u):$(id -g) $HOME/.kube/config

Alternatively, if you are the root user, you can run:

  export KUBECONFIG=/etc/kubernetes/admin.conf

You should now deploy a pod network to the cluster.
Run "kubectl apply -f [podnetwork].yaml" with one of the options listed at:
  https://kubernetes.io/docs/concepts/cluster-administration/addons/

You can now join any number of the control-plane node running the following
command on each as root:

  kubeadm join 192.168.1.7:8443 --token 87z2c8.rxtz78pfculvq5il \
    --discovery-token-ca-cert-hash
sha256:5682f51b25e076b75c7c02139435fed3ebe03b28ae3752599ea3ae6129830969 \
    --control-plane --certificate-key
d87854ac7a180dd5c9720336f6700d76c2e7e998317f7b7e4650d6e90758beff

Please note that the certificate-key gives access to cluster sensitive data,
keep it secret!
As a safeguard, uploaded-certs will be deleted in two hours; If necessary,
you can use
"kubeadm init phase upload-certs --upload-certs" to reload certs afterward.

Then you can join any number of worker nodes by running the following on each
as root:

kubeadm join 192.168.1.7:8443 --token 87z2c8.rxtz78pfculvq5il \
    --discovery-token-ca-cert-hash
sha256:5682f51b25e076b75c7c02139435fed3ebe03b28ae3752599ea3ae6129830969
```

其中--apiserver-advertise-address 选项用来指定本机的 IP 地址，--control-plane-endpoint 选项用来指定 Master 节点控制层的服务地址和端口，实际上就是 keepalived 中配置的虚拟 IP，端口 8443 为 haproxy.cfg 配置文件中配置的前端服务的端口。

初始化成功之后，用户需要留意后面的提示信息，例如用户环境变量的配置以及加入集群的命令。需要注意的是后面有 2 条 kubeadm join 命令，其中一条为其他 Master 节点加入到控制层的命令，另外一条为工作节点加入到集群的命令。

初始化完成之后，查看当前集群中的 Pod，命令如下：

```
[root@master1 ~]# kubectl get pods --all-namespaces
NAMESPACE     NAME                          READY   STATUS   RESTARTS    AGE
kube-system   coredns-7f89b7bc75-8btt4        0/1     Pending        0    80s
kube-system   etcd-master1                    1/1     Running        0    85s
kube-system   kube-apiserver-master1          1/1      Running       0    85s
```

```
kube-system     kube-controller-manager-master1  1/1     Running     0     85s
kube-system     kube-proxy-h7h8q                1/1     Running     0     80s
kube-system     kube-scheduler-master1          1/1     Running     0     85s
```

从上面的输出结果可知，除了 coredns 之外，其余的 Pod 都已经处于运行状态。这是由于还没有安装网络组件，等网络组件安装完成之后，coredns 就正常了。

10.6.8　安装 Calico 网络

Kubernetes 的网络组件非常多，其中比较常用的有 Flannel 和 Calico。而 Calico 是目前稳定性较好，性能也非常高的一种网络组件。因此，在本例中使用 Calico 来进行讲解。以下操作只要在 master1 这个 Master 节点上进行就可以了，不需要在每个 Master 节点上执行。安装 Calico 的命令如下：

```
    [root@master1 ~]# kubectl apply -f https://docs.projectcalico.org/
manifests/calico.yaml
```

安装完成之后，再查看 Pod 状态，如下所示：

```
    [root@master1 ~]# kubectl get pods --all-namespaces
    NAMESPACE     NAME                                     READY   STATUS RESTARTS AGE
    kube-system   calico-kube-controllers-5dc87d545c-txgm4 1/1 Running   0  81m
    kube-system   calico-node-6ld5p                        1/1     Running   0  81m
    kube-system   coredns-7f89b7bc75-8btt4                 1/1     Running   0  93m
    kube-system   etcd-master1                             1/1     Running   0  93m
    kube-system   kube-apiserver-master1                   1/1     Running   0  3m
    kube-system   kube-controller-manager-master1          1/1     Running   2  3m
    kube-system   kube-proxy-h7h8q                         1/1     Running   0  93m
    kube-system   kube-scheduler-master1                   1/1    Running    2  93m
```

可以发现，目前所有的组件都已经处于运行状态。

10.6.9　加入其余的 Master 节点

到目前为止，Kubernetes 集群中已经有了一个 Master 节点，其余的 Master 节点还没有加入到控制层。下面将 master2 和 master3 这 2 个 Master 节点也加入到 Master 节点的控制层中。以下操作分别在 master2 和 master3 上面执行，其中 master2 节点的命令如下：

```
    [root@master2 ~]# kubeadm join 192.168.1.7:8443 --token
87z2c8.rxtz78pfculvq5il \
        --discovery-token-ca-cert-hash
sha256:5682f51b25e076b75c7c02139435fed3ebe03b28ae3752599ea3ae6129830969 \
        --control-plane --certificate-key
d87854ac7a180dd5c9720336f6700d76c2e7e998317f7b7e4650d6e90758beff
```

master3 节点的命令如下：

```
    [root@master3 ~]# kubeadm join 192.168.1.7:8443 --token
87z2c8.rxtz78pfculvq5il \
        --discovery-token-ca-cert-hash
sha256:5682f51b25e076b75c7c02139435fed3ebe03b28ae3752599ea3ae6129830969 \
        --control-plane --certificate-key
d87854ac7a180dd5c9720336f6700d76c2e7e998317f7b7e4650d6e90758beff
```

等待命令执行完成之后，再次查看集群中的 Pod 状态，如下所示：

```
    [root@master1 ~]# kubectl get pods --all-namespaces
    NAMESPACE        NAME                                        READY    STATUS RESTARTS   AGE
    kube-system      calico-kube-controllers-5dc87d545c-txgm4 1/1 Running 0 162m
    kube-system      calico-node-61d5p                           1/1         Running    0  62m
    kube-system      calico-node-99kn6                           1/1         Running    0  162m
    kube-system      calico-node-km8fd                           1/1         Running    0  3m32s
    kube-system      coredns-7f89b7bc75-8btt4                    1/1         Running    0  173m
    ...
```

可以发现，apiserver、controller-manager、etcd 以及 scheduler 等重要组件都已经有 3 个节点在运行。

此时，在 haproxy 的统计报告界面，可以发现 3 个 Master 节点都已经变成绿色了，如图 10-3 所示。

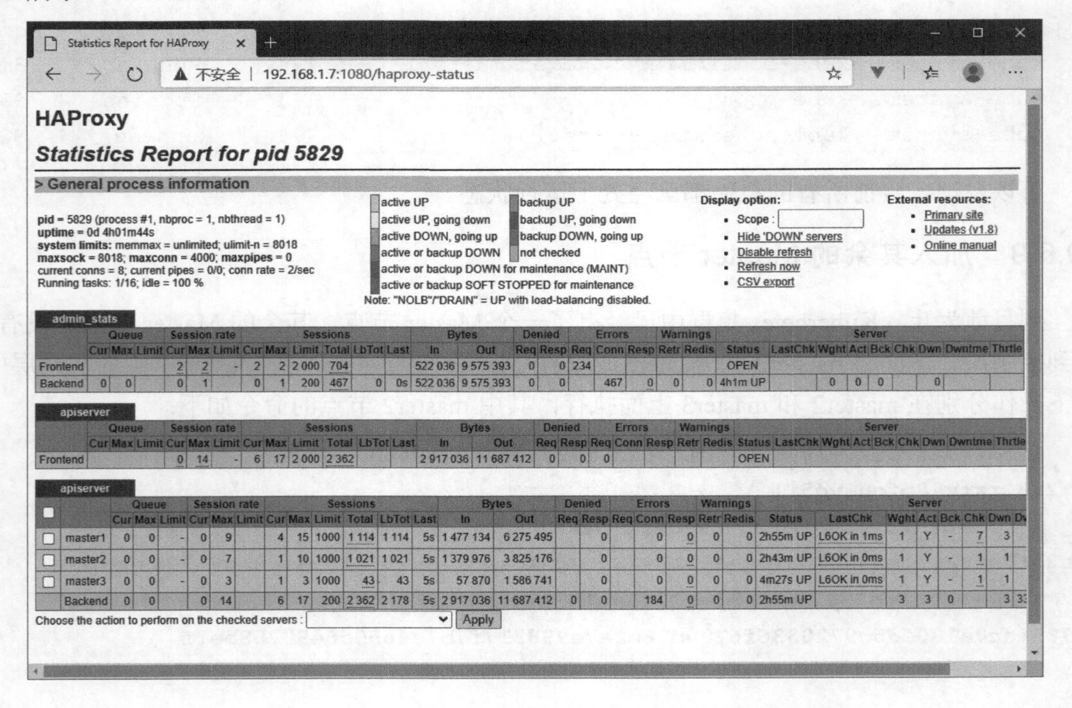

图 10-3　通过 haproxy 统计页面查看 Master 节点状态

10.6.10 加入工作节点

工作节点的加入比较简单，在所有的安装环境中将 kubectl、kubelet 和 kubeadm 等组件都安装完成之后，使用在初始化 master0 时系统给出的命令即可。在本例中，加入命令为：

```
[root@node1 ~]# kubeadm join 192.168.1.7:8443 --token 87z2c8.rxtz78pfculvq5il \
    --discovery-token-ca-cert-hash
sha256:5682f51b25e076b75c7c02139435fed3ebe03b28ae3752599ea3ae6129830969
```

当以上命令执行完成之后，在任何一个 Master 节点上面都可以使用以下命令查看当前集群中的工作节点列表：

```
[root@master1 ~]# kubectl get node
NAME      STATUS   ROLES                  AGE    VERSION
master1   Ready    control-plane,master   5h11m  v1.20.0
master2   Ready    control-plane,master   5h     v1.20.0
master3   Ready    control-plane,master   141m   v1.20.0
node1     Ready    <none>                 62s    v1.20.0
```

从上面的输出结果可知，node1 已经处于 Ready（就绪）状态了。

第11章

实战1：在Kubernetes集群中部署Spring Boot应用程序

Spring Boot 是目前最为流行的 Web 开发框架之一。通过 Spring Boot，用户可以快速开发自己的应用系统。本章将以一个具体的应用系统为例，介绍如何在 Kubernetes 集群中部署 Spring Boot 应用系统。

11.1 应用系统概况

通常情况下，用户的应用系统都是已经开发完成的。在本书中，为了便于讲解，我们准备了一个非常简单的 Spring Boot 应用系统，该系统的功能比较单一，就是把 MySQL 数据库中的城市数据查询出来，以表格的形式显示在网页上面。关于该应用系统的代码，请参照附件。

由于演示的应用系统涉及 MySQL 数据库操作，因此还需要将所使用的 MySQL 数据库导出为 SQL 脚本文件，以便于在 MySQL 容器中恢复数据。在本例中，SQL 脚本文件为 world.sql。

11.2 部署 MySQL

首先为 MySQL 创建一个单独的命名空间，YAML 文件的代码如下：

```
01  apiVersion: v1
02  kind: Namespace
03  metadata:
04    name: mysql-ns
05    labels:
06      name: mysql-ns
```

其中第 4 行指定所创建的命名空间的名称为 mysql-ns，将以上文件保存为 mysql-ns.yaml，然后使用以下命令创建命名空间：

```
[root@master1 ~]# kubectl apply -f mysql-ns.yaml
namespace/mysql-ns created
```

创建完成之后，查看命名空间状态是否正常，如下所示：

```
[root@master1 ~]# kubectl get namespaces | grep mysql
mysql-ns              Active          4m6s
```

从上面的输出结果可知，所创建的命名空间已经处于活动状态。

接下来，为 MySQL 创建一个类型为 StatefulSet 的资源，其 YAML 文件代码如下：

```
01 apiVersion: apps/v1
02 kind: StatefulSet
03 metadata:
04   name: mysql
05 spec:
06   selector:
07     matchLabels:
08       app: mysql
09   serviceName: mysql
10   replicas: 1
11   template:
12     metadata:
13       labels:
14         app: mysql
15     spec:
16       containers:
17       - name: mysql
18         image: mysql/mysql-server
19         env:
20         - name: MYSQL_ALLOW_EMPTY_PASSWORD
21           value: "1"
22         ports:
23         - name: mysql
24           containerPort: 3306
25           protocol: TCP
```

第 9 行指定解析该 Pod 的 IP 地址的服务的名称。第 18 行指定所使用的镜像为 mysql/mysql-server。第 19 行通过 env 字段为 MySQL 指定一些环境变量，在本例中，设置的环境变量为 MYSQL_ALLOW_EMPTY_PASSWORD，其值为 1，表示允许空密码。第 24 行指定容器暴露的端口为 3306。

将以上代码保存为 mysql-statefullset.yaml，然后使用以下命令创建资源：

```
[root@master1 ~]# kubectl apply -f appservice.yaml --namespace=mysql-ns
```

创建完成之后，查看 Pod 状态，命令如下：

```
[root@master1 ~]# kubectl get pods --namespace=mysql-ns
NAME         READY   STATUS        RESTARTS         AGE
```

```
mysql-0     1/1     Running     0                   135m
```

从上面的输出结果可知，MySQL 容器已经成功运行。

再为 MySQL 容器创建一个服务，便于外部访问，其 YAML 代码如下：

```
01  apiVersion: v1
02  kind: Service
03  metadata:
04    name: mysql
05    labels:
06      app: mysql
07  spec:
08    type: NodePort
09    ports:
10    - port: 3306
11      name: mysql
12      targetPort: 3306
13      nodePort: 30006
14    selector:
15      app: mysql
```

第 12 行指定目标端口为 3306，这个端口是前面创建 MySQL 容器时所暴露的服务端口。第 13 行指定 nodePort 为 30006，外部可以通过该端口访问到 MySQL 服务。第 15 行指定标签为 mysql 的 Pod 后端服务，这个标签也是前面创建 MySQL 服务时设置的。

将以上代码保存为 mysql-service.yaml，然后使用以下命令创建资源：

```
[root@master1 ~]# kubectl apply -f mysql-service.yaml --namespace=mysql-ns
```

创建完成之后，用户原则上就可以通过 MySQL Pod 所在的节点的 IP 地址和 nodePort 端口访问 MySQL 服务器了。但是实际上用户连接 MySQL 服务器时通常会出现以下错误：

```
[root@master1 ~]# mysql -h 192.168.2.131 -uroot -P 30006
ERROR 1045 (28000): Access denied for user 'root'@'192.168.2.131' (using
password: NO)
```

以上错误出现的原因是当前 MySQL 服务器不允许 root 用户远程访问，用户需要另外创建一个用户。创建 MySQL 用户的方法如下。

首先查看 MySQL Pod 所在的节点，如下所示：

```
[root@master1 ~]# kubectl get pods --namespace=mysql-ns -o wide
NAME       READY   STATUS      RESTARTS   AGE    IP              NODE
NOMINATED  NODE    READINESS  GATES
mysql-0    1/1     Running     0          157m   10.122.166.134  node1
<none>  <none>
```

从上面的输出结果可知，MySQL 所在的 Pod 位于名称为 node1 的节点上。

然后在 node1 节点上面查看容器列表，如下所示：

```
[root@node1 demo]# docker ps
CONTAINER ID  IMAGE                COMMAND                     CREATED STATUS
PORTS   NAMES
391d9c11ce92   mysql/mysql-server  "/entrypoint.sh mysq…"  3 hours ago Up 3
hours
k8s_mysql_mysql-0_mysql-ns_a976968b-b120-46c1-ad0c-0281a0db44e7_0
…
```

从上面的输出结果可知，MySQL 的容器 ID 为 391d9c11ce92，使用以下命令进入到该容器：

```
[root@node1 ~]# docker exec -it 391d9c11ce92 bash
```

其中 391d9c11ce92 为容器 ID，bash 为进入容器内部之后所要执行的命令。执行完成之后，可以发现命令提示符已经发生变化，然后使用 mysql 命令连接到 MySQL 服务器，如下所示：

```
bash-4.2# mysql -uroot
Welcome to the MySQL monitor.  Commands end with ; or \g.
Your MySQL connection id is 47
Server version: 8.0.22 MySQL Community Server - GPL

Copyright (c) 2000, 2020, Oracle and/or its affiliates. All rights reserved.

Oracle is a registered trademark of Oracle Corporation and/or its
affiliates. Other names may be trademarks of their respective
owners.

Type 'help;' or '\h' for help. Type '\c' to clear the current input statement.

mysql>
```

为应用系统创建一个 MySQL 连接用户，命令如下：

```
mysql> CREATE USER web@'%' IDENTIFIED WITH MYSQL_NATIVE_PASSWORD BY
'Hawk123!';
```

命令中 web@'%'为用户名，其中的%表示该用户可以从任何主机连接 MySQL 服务器。后面的 Hawk123!为密码。

然后为 web@'%'授予访问数据库的权限，如下所示：

```
mysql> GRANT ALL PRIVILEGES ON *.* TO web@'%';
```

此时，用户就可以使用 web 用户名来连接 MySQL 服务器了。图 11-1 所示显示的是 Navicat Premium 15 的连接选项。

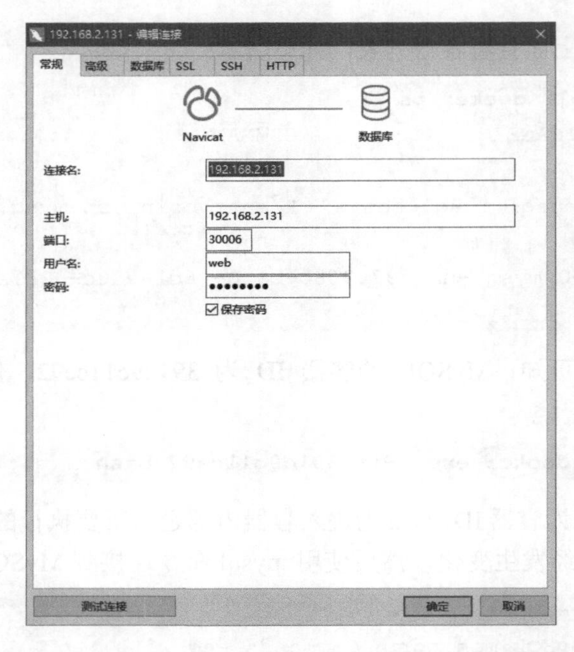

图 11-1　通过 Navicat Premium 15 连接 MySQL 服务器

连接成功之后，创建一个名称为world的数据库，然后使用前面导出的数据库脚本world.sql创建数据表和恢复数据。

11.3　准备应用系统

修改 application.properties 文件的内容，将其中数据库地址中的 IP 地址修改为 node1 的 IP 地址，端口为 nodePort 的值，用户名和密码分别为前面所创建的值，如下所示：

```
    01  spring.datasource.url=jdbc:mysql://192.168.21.131:30006/
world?serverTimezone=GMT
    02  spring.datasource.username=web
    03  spring.datasource.password=Hawk123!
    04  spring.datasource.driver-class-name=com.mysql.cj.jdbc.Driver
    05  spring.datasource.max-idle=10
    06  spring.datasource.max-wait=10000
    07  spring.datasource.min-idle=5
    08  spring.datasource.initial-size=5
    09  spring.jpa.database-platform=org.hibernate.dialect.MySQLDialect
    10  spring.thymeleaf.mode=HTML5
    11  spring.thymeleaf.encoding=UTF-8
    12  spring.thymeleaf.content-type=text/html
    13  spring.thymeleaf.cache=false
    14  logging.level.root=debug
    15  logging.file.path=/
```

为了便于部署，用户需要将其打包成为一个 jar 文件。打包的方法使用 Maven 插件。如果用户的应用是使用 IntelliJ IDEA 开发的，则该工具已经内置了 Maven 插件，如图 11-2 所示。

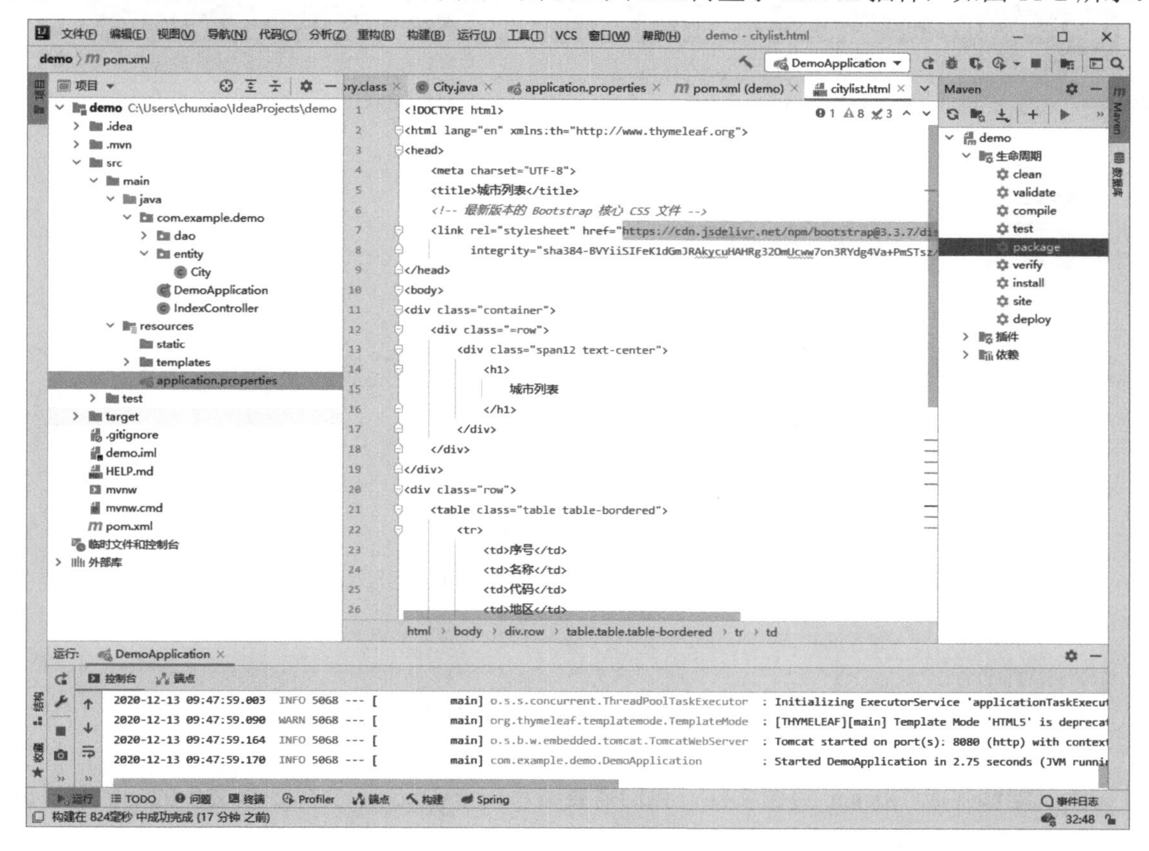

图 11-2　IntelliJ IDEA

在 IntelliJ IDEA 的右侧打开 Maven 面板，双击其中的 package 命令，在底部的“运行”窗口中会不断输出日志信息，如下所示：

```
    [INFO] --- maven-jar-plugin:3.2.0:jar (default-jar) @ demo ---
    [INFO] Building jar:
C:\Users\chunxiao\IdeaProjects\demo\target\demo-0.0.1-SNAPSHOT.jar
    [INFO]
    [INFO] --- spring-boot-maven-plugin:2.4.1:repackage (repackage) @ demo ---
    [INFO] Replacing main artifact with repackaged archive
    [INFO] ------------------------------------------------------------------------
    [INFO] BUILD SUCCESS
    [INFO] ------------------------------------------------------------------------
    [INFO] Total time:  11.217 s
    [INFO] Finished at: 2020-12-13T10:26:10+08:00
    [INFO] ------------------------------------------------------------------------
```

当出现 BUILD SUCCESS 之后，表示已经打包完成。

打包生成的 jar 文件位于项目目录中的 target 目录中，如图 11-3 所示。

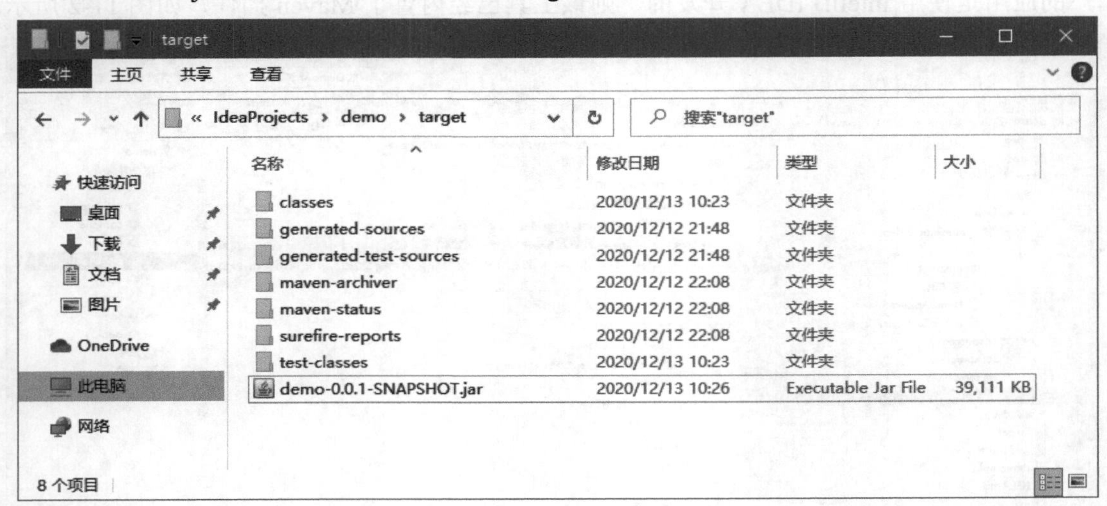

图 11-3　打包生成的 jar 文件

在后面的步骤中，我们需要把生成的名称为 demo-0.0.1-SNAPSHOT.jar 的文件放到一个 Docker 镜像文件中。

11.4　编写 Docker 文件

为了能够生成 Docker 镜像文件，我们需要编写一个 Docker 文件。该文件的功能是告诉 Docker 如何构建镜像文件，代码如下：

```
01  FROM java:8
02  VOLUME /tmp
03  ADD demo-0.0.1-SNAPSHOT.jar /app.jar
04  RUN sh -c 'touch /app.jar'
05  ENV JAVA_OPTS=""
06  ENTRYPOINT [ "sh", "-c", "java $JAVA_OPTS 
-Djava.security.egd=file:/dev/./urandom -jar /app.jar" ]
```

第 1 行的 FROM 指令表明该 Docker 镜像是在 java:8 这个镜像的基础上构建的。FROM 指令的功能是用来指定基础镜像。第 2 行的 VOLUME 指令的功能是定义匿名数据卷，在本例中，VOLUME 指令指向/tmp。第 3 行的 ADD 指令的功能是从上下文目录中复制文件或者目录到容器里指定路径。在本例中是将前面打包生成的 demo-0.0.1-SNAPSHOT.jar 文件，复制到容器根目录中，并且命名为 app.jar。第 4 行的 RUN 指令用来执行后面的命令。第 5 行的 ENV 指令用来配置环境变量，在本例中不需要额外指定环境变量。第 6 行的 ENTRYPOINT 指令用来指定容器启动时需要执行的命令。在本例中，其功能是将前面打包生成的 jar 文件在容器启动时一起启动。

11.5 构建镜像

接下来，需要将用户的应用系统构建在一个镜像文件中。创建一个新的目录，把前面生成的 demo-0.0.1-SNAPSHOT.jar 文件和 Docker 文件都放在该目录中，注意 Docker 文件的文件名为 Dockerfile。最后目录的内容如下：

```
[root@master2 demo]# ll
total 39116
-rw-r--r-- 1   root    root        40049005    Dec 13 10:26
demo-0.0.1-SNAPSHOT.jar
-rw-r--r-- 1   root    root        204         Dec 13 10:57
Dockerfile
```

然后使用 docker build 命令构建镜像，如下所示：

```
[root@master2 demo]# docker build -t demo .
Sending build context to Docker daemon  40.05MB
Step 1/6 : FROM java:8
8: Pulling from library/java
5040bd298390: Pull complete
fce5728aad85: Pull complete
76610ec20bf5: Pull complete
60170fec2151: Pull complete
e98f73de8f0d: Pull complete
11f7af24ed9c: Pull complete
49e2d6393f32: Pull complete
bb9cdec9c7f3: Pull complete
Digest:
sha256:c1ff613e8ba25833d2e1940da0940c3824f03f802c449f3d1815a66b7f8c0e9d
Status: Downloaded newer image for java:8
 ---> d23bdf5b1b1b
Step 2/6 : VOLUME /tmp
 ---> Running in 76bd7a4bcd1b
Removing intermediate container 76bd7a4bcd1b
 ---> 792223e07fdc
Step 3/6 : ADD demo-0.0.1-SNAPSHOT.jar /app.jar
 ---> 6792cb1bbdc2
Step 4/6 : RUN sh -c 'touch /app.jar'
 ---> Running in bf670261fb8d
Removing intermediate container bf670261fb8d
 ---> 8ef67e780bdd
```

```
    Step 5/6 : ENV JAVA_OPTS=""
     ---> Running in fb1d3e7621e9
    Removing intermediate container fb1d3e7621e9
     ---> a30b30eaff9b
    Step 6/6 : ENTRYPOINT [ "sh", "-c", "java $JAVA_OPTS
-Djava.security.egd=file:/dev/./urandom -jar /app.jar" ]
     ---> Running in a09aecba4258
    Removing intermediate container a09aecba4258
     ---> d184838b98db
    Successfully built d184838b98db
    Successfully tagged demo:latest
```

其中-t 选项用来指定新的镜像的名称和标签，其格式为 name:tag 或者 name，在本例中，只指定了镜像名称，没有指定标签。命令最后的圆点表示当前目录。

构建完成之后，查看当前系统中的镜像列表，如下所示：

```
    [root@master2 demo]# docker images
    REPOSITORY  TAG              IMAGE ID       CREATED          SIZE
    demo        latest           d184838b98db   11 minutes ago   723MB
    …
```

可以发现，刚刚构建的镜像已经出现在列表中。

11.6 部署应用系统

接下来开始部署用户的 Spring Boot 应用系统。首先需要创建一个 Deployment 资源对象，其代码如下：

```
01  apiVersion: apps/v1
02  kind: Deployment
03  metadata:
04    name: demo
05  spec:
06    replicas: 1
07    selector:
08      matchLabels:
09        app: demo
10
11    template:
12      metadata:
13        labels:
14          app: demo
15      spec:
```

```
16        containers:
17          - name: demo
18            image: demo
19            imagePullPolicy: Never
20            ports:
21              - containerPort: 8080
```

第 18 行通过 image 字段指定容器使用的镜像为 demo，这个镜像文件是前面构建的。第 19 行通过 imagePullPolicy 字段设置不从远程软件仓库拉取镜像。

将以上代码保存为 demo-deployment.yaml，然后使用以下命令进行部署：

```
[root@master1 ~]# kubectl apply -f demo-deployment.yaml
```

查看 Pod 状态，如下所示：

```
[root@master1 ~]# kubectl get pods
NAME                      READY       STATUS      RESTARTS        AGE
demo-6dbd5f6475-pfxdx     1/1         Running     0               97m
…
```

可以得知所创建的 Pod 已经处于运行状态了。

为了便于访问，下面再为应用系统创建一个 nodePort 类型的服务，其 YAML 配置文件如下：

```
01 apiVersion: v1
02 kind: Service
03 metadata:
04   name: appdemo
05   labels:
06     app: appdemo
07 spec:
08   type: NodePort
09   ports:
10   - port: 8080
11     name: demoapp
12     targetPort: 8080
13     nodePort: 30080
14   selector:
15     app: demo
```

将以上代码保存为 demo-service.yaml，然后使用以下命令创建服务：

```
[root@master1 ~]# kubectl apply -f demo-service.yaml
```

最后，用户就可以通过该服务访问应用系统了，如图 11-4 所示。

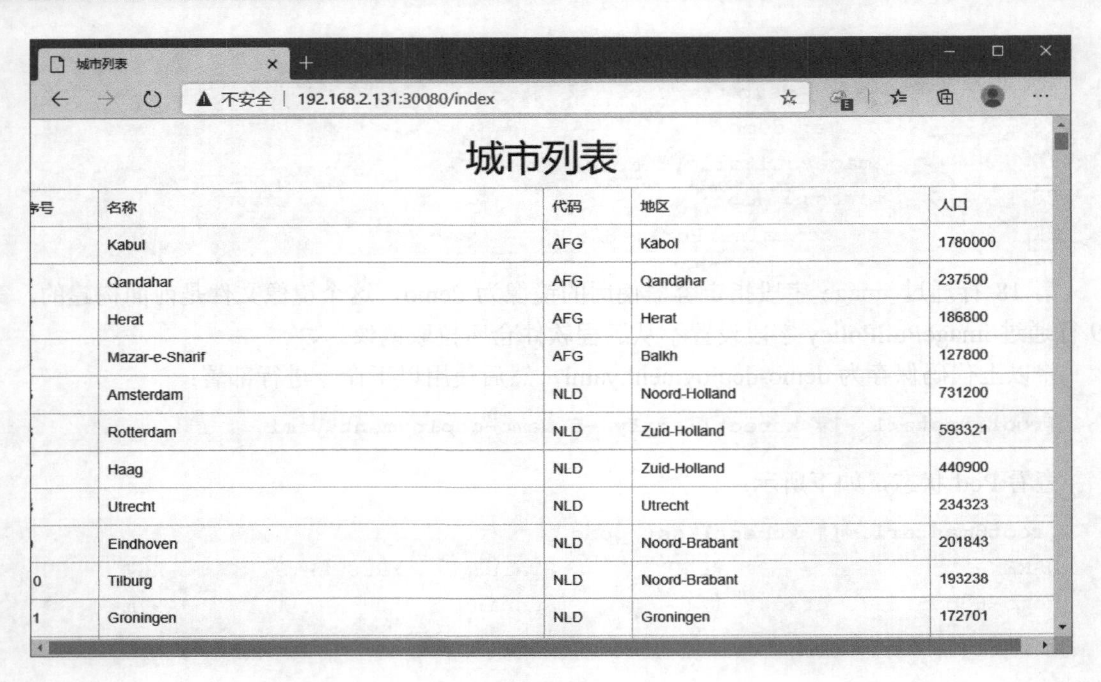

城市列表

序号	名称	代码	地区	人口
	Kabul	AFG	Kabol	1780000
	Qandahar	AFG	Qandahar	237500
	Herat	AFG	Herat	186800
	Mazar-e-Sharif	AFG	Balkh	127800
	Amsterdam	NLD	Noord-Holland	731200
	Rotterdam	NLD	Zuid-Holland	593321
	Haag	NLD	Zuid-Holland	440900
	Utrecht	NLD	Utrecht	234323
	Eindhoven	NLD	Noord-Brabant	201843
0	Tilburg	NLD	Noord-Brabant	193238
1	Groningen	NLD	Groningen	172701

图 11-4　应用系统界面

第 12 章

实战2：安装KubeSphere

在前面的所有例子中，各种资源基本上都是通过 YAML 配置文件创建的。如果不熟悉资源的元数据，很难创建出合适的资源。尽管 Dashboard 给用户提供了一定的便利性，但是其功能仍然比较单一。

如果将 Kubernetes 比作是 Linux 内核，那么 KubeSphere 则是各种 Linux 发行版。它为 Kubernetes 提供了极大的易操作性。KubeSphere 愿景是打造一个以 Kubernetes 为内核的云原生分布式操作系统，它的架构可以非常方便地使第三方应用与云原生生态组件进行即插即用地集成，支持云原生应用在多云与多集群的统一分发和运维管理。它提供全栈的 IT 自动化运维的能力，简化企业的 DevOps 工作流。本章将详细介绍 KubeSphere 的安装和使用方法。

12.1 安装 KubeSphere

KubeSphere 为用户提供了非常方便的安装方式，主要有 All-in-one 和在已有集群上面安装两种方式。本节将详细介绍这两种安装方法。

12.1.1 安装条件

KubeSphere 目前最新的版本为 3.0，其官方网站为：https://kubesphere.io/。

其中主界面如图 12-1 所示。

KubeSphere 官方为用户提供了两种安装方式，其中一种是在已有的 Kubernetes 集群上安装，另外一种是非常方便的 All-in-one 安装方式。如果用户是在生产环境中使用 KubeSphere，需要使用第 1 种安装方式。如果用户是一个初学者，想要快速体验 KubeSphere，则可以使用第 2 种安装方式。

对于第 1 种安装方式，Kubernetes 的版本必须是 1.15.0 以上，硬件要求至少保证 CPU 有 1 个内核，内存不少于 2GB。另外在 Kubernetes 集群中必须预先配置了存储类，apiserver 的 CSR 功能也必须激活。

对于第 2 种安装方式，要求操作系统为 Ubuntu 16.04 或者 18.04、Debian Buster 或者 Stretch、CentOS 7.x 以及 Red Hat Enterprise Linux 7 等，关于硬件的详细需求，用户可以参考其官方网站。

图 12-1　KubeSphere 主页面

12.1.2　All-in-one 安装

All-in-one 安装方式为初学者提供了一种一键安装 Kubernetes 和 KubeSphere 的途径。在这种方式下，用户只需要一个 kk 命令，即可在一台机器上面完成 Kubernetes 和 KubeSphere 的安装，不需过多的人工干预。

KubeSphere 的主要管理工具为 KubeKey，用户需要首先下载该工具，然后使用该工具安装 KubeSphere。

（1）KubeKey 的下载方法为：

```
[root@kubesphere ~]# curl -sfL https://get-kk.kubesphere.io | VERSION=v1.0.1
sh -
Downloading kubekey v1.0.1 from https://kubernetes.pek3b.qingstor.com/
kubekey/releases/download/v1.0.1/kubekey-v1.0.1-linux-amd64.tar.gz ...
Kubekey v1.0.1 Download Complete!
```

对于国内用户，KubeSphere 专门提供了镜像站点，以加快安装速度。用户在执行安装之前，需要设置 KKZONE 环境变量，如下所示：

```
[root@kubesphere ~]# export KKZONE=cn
```

然后再使用前面的命令下载 KubeKey。

（2）当下载完成之后，会在当前目录中生成一个名称为 kk 的可执行文件。如果该文件没有执行权限，则用户需要使用以下命令进行授权：

```
[root@kubesphere ~]# chmod +x kk
```

（3）接下来的操作主要是通过 kk 管理工具进行的，安装 KubeSphere 的命令如下：

```
[root@kubesphere ~]# ./kk create cluster --with-kubernetes v1.17.9
--with-kubesphere v3.0.0
```

其中--with-kubernetes 选项用来指定 Kubernetes 的版本，--with-kubesphere 选项用来指定 KubeSphere 的版本。

（4）接下来就是等待安装完成，这个过程花费的时间比较长，视用户的网络情况而定，用户需要耐心等待。安装过程中的每个步骤 KubeKey 都给出了详细的日志，用户可以通过观察日志来了解安装进度。如果安装过程出现错误，用户也可以通过日志来分析原因。

（5）安装完成之后，KubeKey 会给出一个验证是否安装成功的命令，如下所示：

```
[root@kubesphere ~]# kubectl logs -n kubesphere-system $(kubectl get pod -n
kubesphere-system -l app=ks-install -o jsonpath='{.items[0].metadata.name}') -f
```

如果安装成功，则以上命令会在最后输出以下信息：

```
#####################################################
###              Welcome to KubeSphere!           ###
#####################################################

Console: http://192.168.1.10:30880
Account: admin
Password: P@88w0rd

NOTES：
  1. After logging into the console, please check the
     monitoring status of service components in
     the "Cluster Management". If any service is not
     ready, please wait patiently until all components
     are ready.
  2. Please modify the default password after login.

#####################################################
https://kubesphere.io             2020-12-15 08:24:35
#####################################################
```

以上信息的输出表示 KubeSphere 已经安装成功，并且给出了控制台的地址、账号和默认的密码。用户可以通过以上信息登录控制台。

注　意

目前 KubeSphere 3.0 对于 CentOS 8 的安装支持还不是很好，建议另外搭建虚拟机环境，在 CentOS 7 上面体验 KubeSphere。

12.1.3 在已有集群上安装 KubeSphere

如果用户已经安装好 Kubernetes 集群，则可以使用 YAML 配置文件来安装 KubeSphere。安装方法也比较简单，命令如下所示：

```
[root@kubesphere ~]# kubectl apply -f https://github.com/kubesphere/
ks-installer/releases/download/v3.0.0/kubesphere-installer.yaml
[root@kubesphere ~]# kubectl apply -f https://github.com/kubesphere/
ks-installer/releases/download/v3.0.0/cluster-configuration.yaml
```

当安装完成之后，使用以下命令查看安装日志，以检验是否安装成功：

```
[root@kubesphere ~]# kubectl logs -n kubesphere-system $(kubectl get pod -n
kubesphere-system -l app=ks-install -o jsonpath='{.items[0].metadata.name}') -f
```

使用以下命令查看 Pod 状态：

```
[root@kubesphere ~]# kubectl get pod --all-namespaces
```

12.2 通过 KubeSphere 管理集群

KubeSphere 为用户提供了基于图形界面的集群管理方式，通过 KubeSphere 的控制台，用户可以查看各个服务组件的状态，管理部署、服务以及项目等资源，本节将详细介绍 KubeSphere 的使用方法。

12.2.1 登录 KubeSphere 控制台

在浏览器中输入前面给出的控制台地址，出现登录界面，如图 12-2 所示。

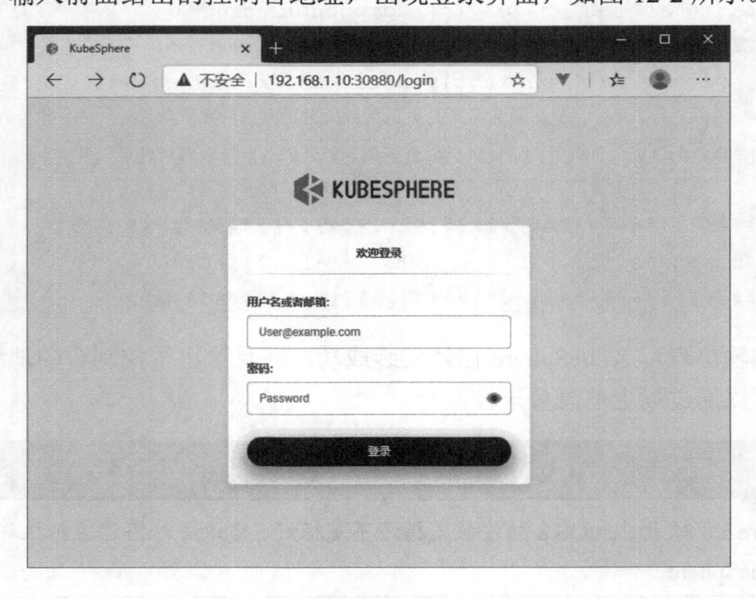

图 12-2 登录 KubeSphere 控制台

输入前面给出的账号和密码，登录进入控制台，出现仪表盘，如图 12-3 所示。

图 12-3　KubeSphere 仪表盘

点击“平台管理”链接，出现管理平台界面，如图 12-4 所示。

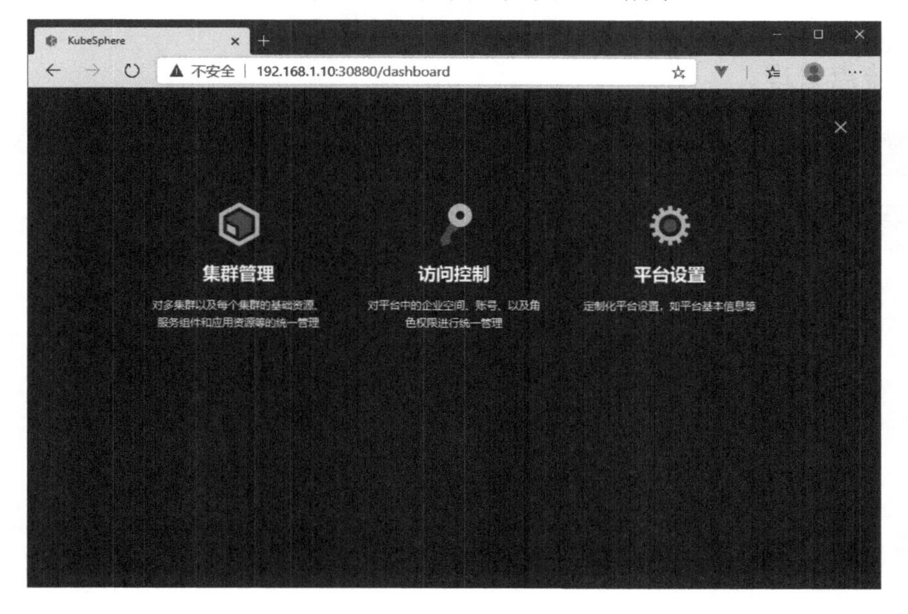

图 12-4　管理平台

图 12-4 一共有 3 个菜单，其中最常用的是“集群管理”。点击“集群管理”链接，打开集群管理界面，如图 12-5 所示。

在图 12-5 中，左侧为功能菜单，一共包括概览、节点管理、服务组件、项目管理、应用负载、配置中心、自定义资源 CRD、存储管理、监控告警以及集群设置等项目。

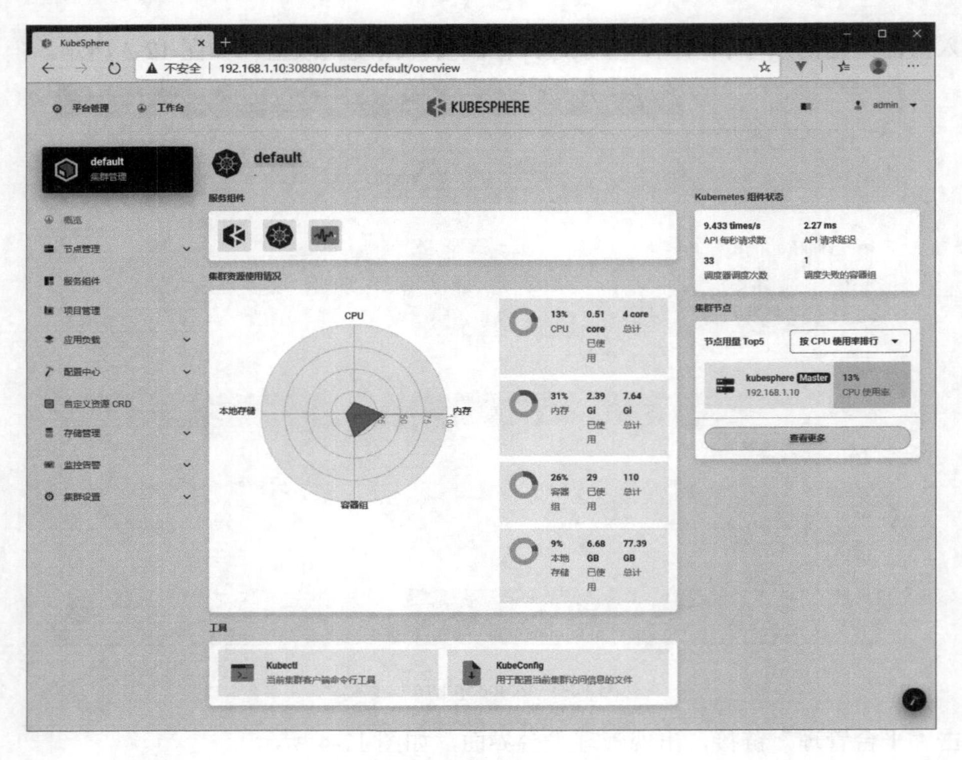

图 12-5　集群管理界面

12.2.2　节点管理

节点管理显示了当前集群中的各种节点的状态，如图 12-6 所示。

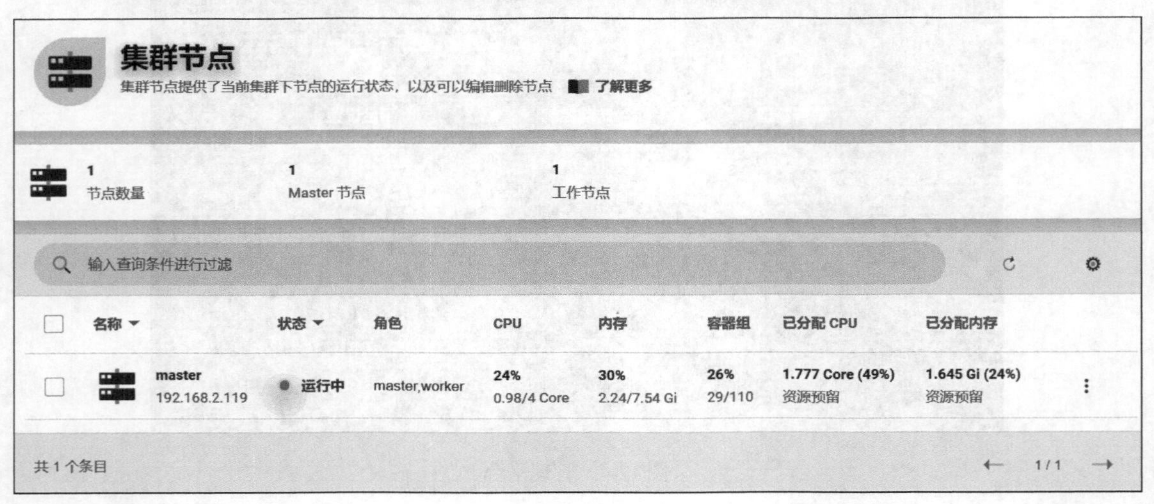

图 12-6　节点状态

从图 12-6 可知，当前集群中有 1 个节点，其中 1 个 Master 节点，1 个工作节点。下面的节点列表每 1 行描述了一个节点，点击右侧的 ⋮ 按钮，可以对该节点进行设置是否可以调度。

12.2.3　服务组件状态查看

服务组件模块显示了当前 KubeSphere 系统中的各种服务组件的状态，如图 12-7 所示。

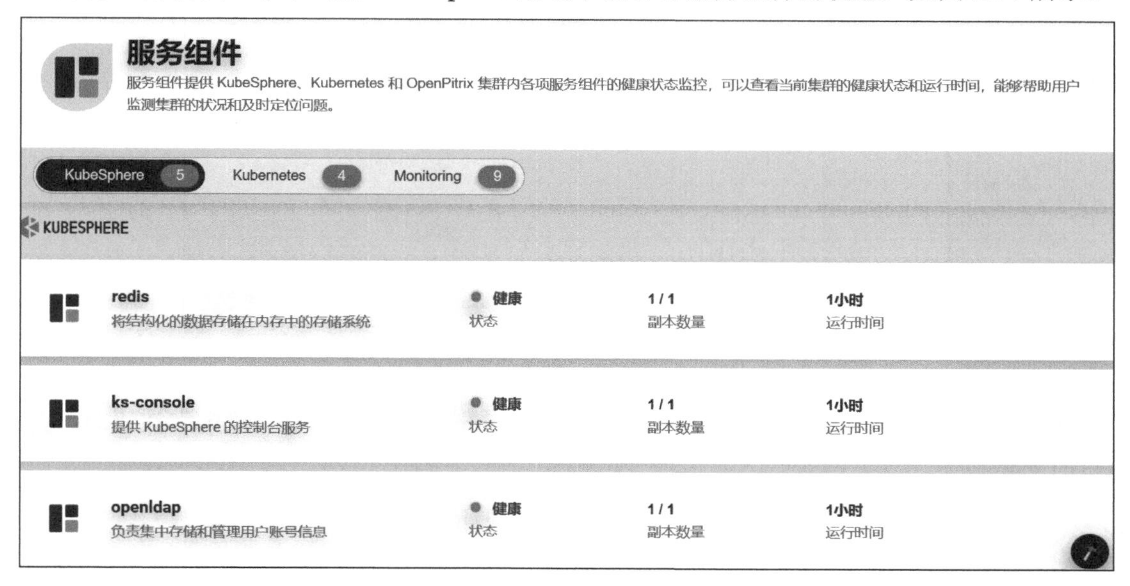

图 12-7　服务组件

所有的服务组件分为 3 个部分，其中 KubeSphere 有 5 个，Kubeletes 有 4 个，Monitoring 有 9 个。

12.2.4　项目管理

在 KubeSphere 中，项目对应 Kubeletes 的命名空间，如图 12-8 所示。

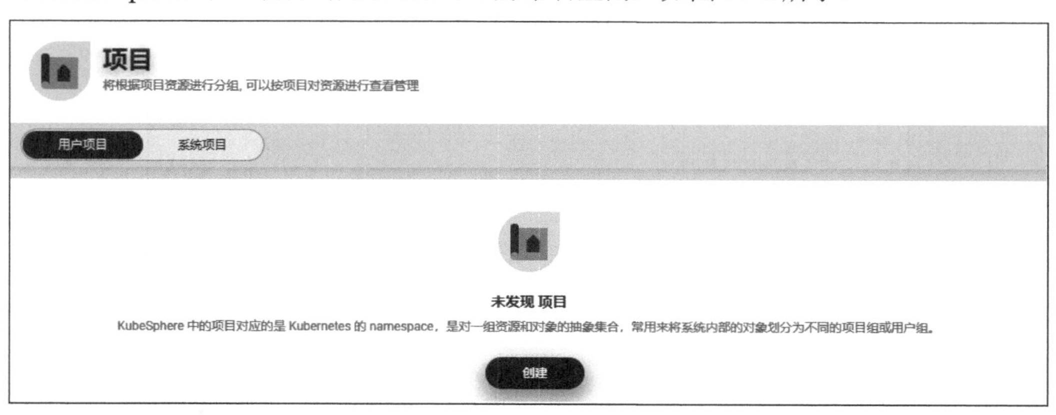

图 12-8　项目管理

项目分为用户项目和系统项目，其中用户项目为用户自己创建的命名空间，而系统项目则是 KubeSphere 系统本身的命名空间。点击用户项目页面底部的“创建”按钮，可以创建一个新的项目，如图 12-9 所示。

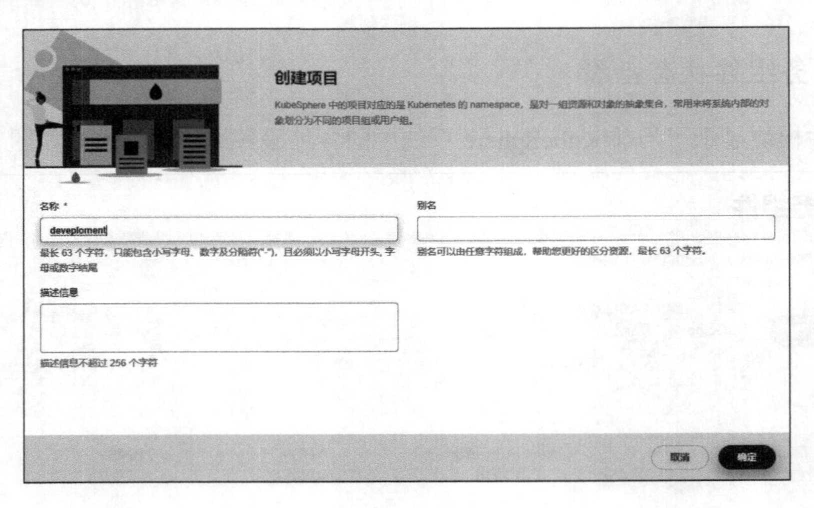

图 12-9　创建项目

在“名称”文本框中输入项目名称，例如 development，然后单击“确定”按钮即可完成项目的创建。

12.2.5　工作负载管理

工作负载模块是最重要的模块，在该模块中用户可以管理部署、任务、服务和应用路由等 Kubeletes 资源。下面演示如何使用 KubeSphere 控制台创建一个 Nginx 应用。

（1）打开在工作负载界面，在“部署”标签页中点击右侧的“创建”按钮。打开“创建部署”对话框，如图 12-10 所示。

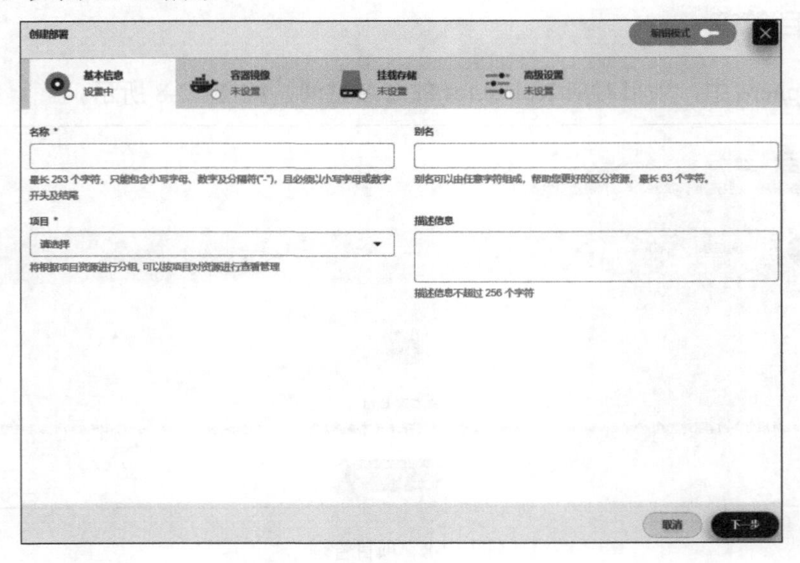

图 12-10　“创建部署”对话框

在“名称”文本框中输入部署的名称，例如 webserver，点击“项目”下拉菜单，选择前面创建的 development 项目，点击“下一步”按钮，进入“容器镜像”对话框，如图 12-11 所示。

图 12-11　“容器镜像”对话框

（2）容器组副本数量即前面介绍的 replicas 选项，用户可以根据自己的实际需要进行设置。点击“添加容器镜像”按钮，打开“添加容器”对话框，如图 12-12 所示。

图 12-12　“添加容器”对话框

（3）容器设置。在图 12-12 所示的界面中，点击 DockerHub 右侧的 按钮，进行镜像搜索。输入 nginx，然后在下拉菜单中选择官方的 nginx 镜像，如图 12-13 所示。

图 12-13　选择 nginx 镜像

此外，用户还可以进行其他的容器设置，例如容器名称、容器类型、资源限制以及端口设置等，用户可以根据自己的实际情况进行修改，如图 12-14 所示。

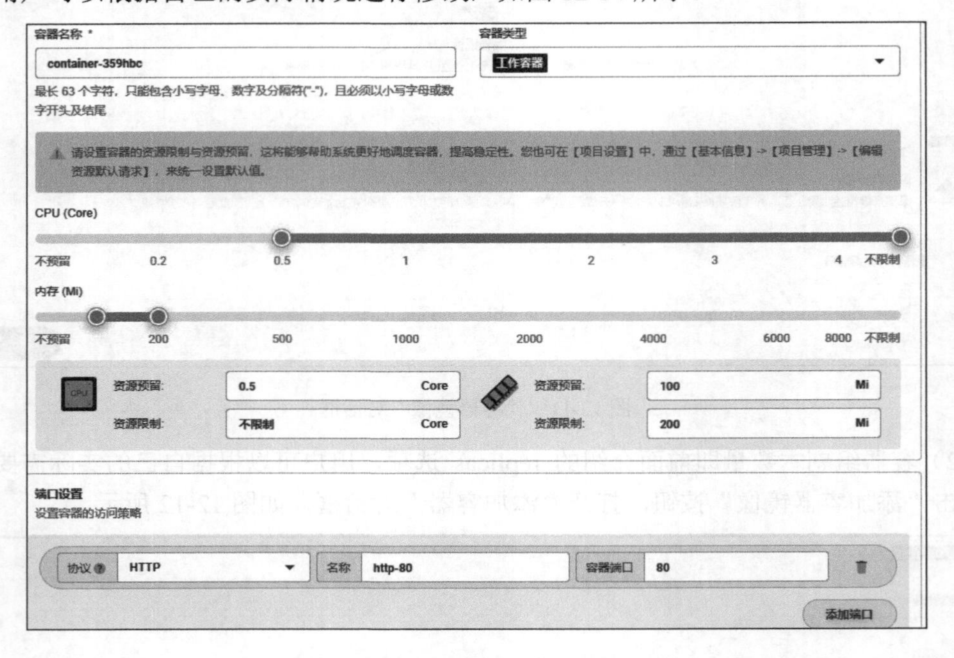

图 12-14　配置容器

配置完成之后，点击底部的按钮，关闭“添加容器”对话框。继续点击“下一步”按钮，进行剩下的操作。

（4）设置存储卷。接下来是添加存储卷，如图 12-15 所示。

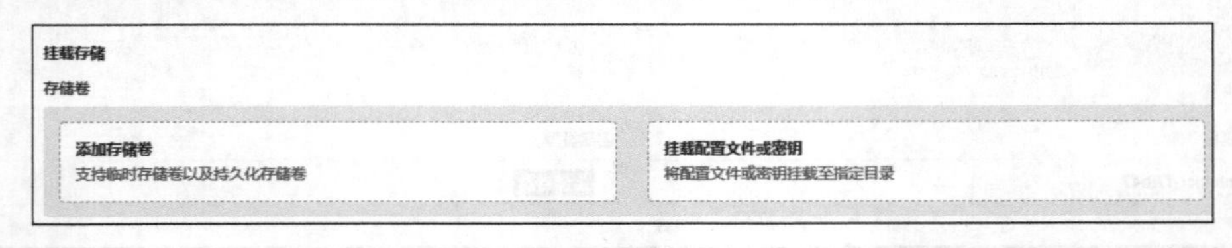

图 12-15　“存储卷”管理

如果用户需要为 Nginx 应用添加存储卷，可以点击“添加存储卷”按钮进行添加，如果没有需要，则可以直接点击“下一步”按钮。

（5）高级设置。在该界面中，用户可以设置节点的调度策略以及标签等元数据，如图 12-16 所示。如果没有需要，则点击“创建”按钮，完成 Nginx 应用的部署。

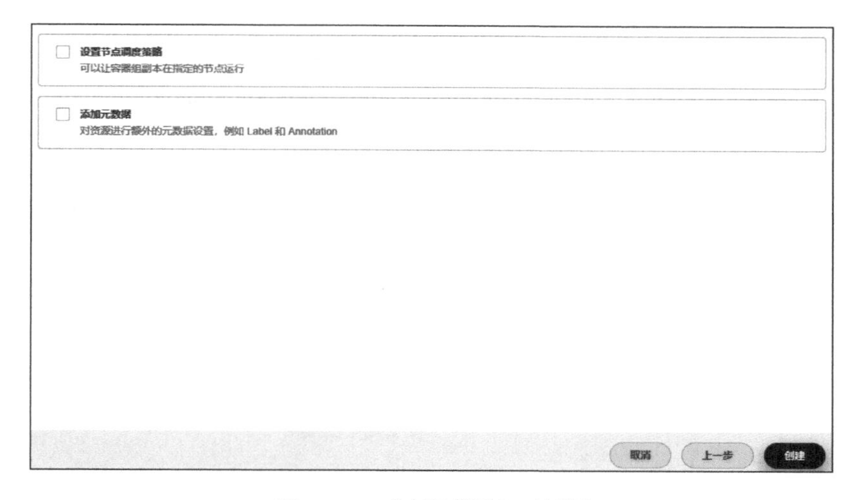

图 12-16　“高级设置”对话框

稍等片刻之后，在部署列表中就可以看到新创建的 Nginx 应用处于运行中了，如图 12-17 所示。

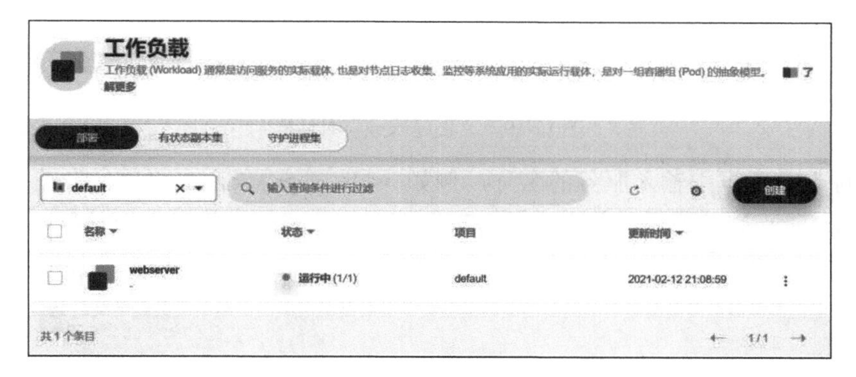

图 12-17　查看部署状态

12.2.6　服务管理

服务模块包含了当前 Kubeletes 集群中的各种服务清单，如图 12-18 所示。

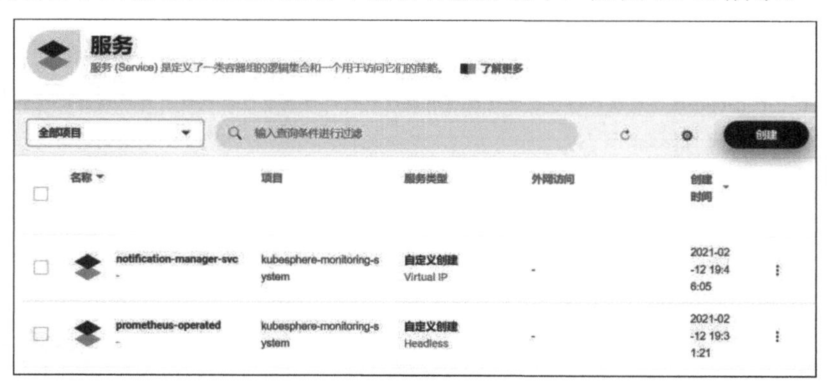

图 12-18　服务管理

接下来介绍如何创建一个服务，使得前面创建的 Nginx 可以被外部访问。

（1）单击右侧的“创建”按钮，打开“创建服务”对话框，如图 12-19 所示。

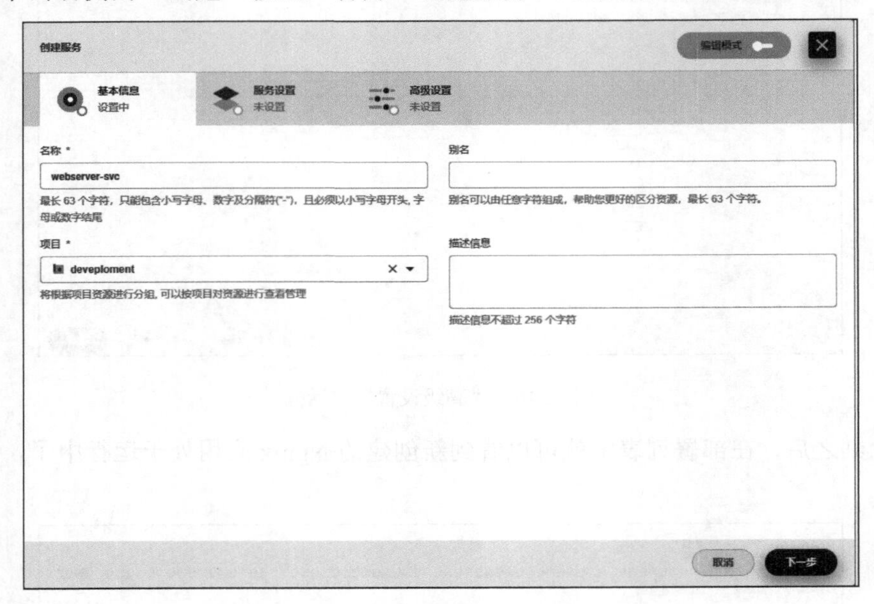

图 12-19 “创建服务”对话框

在“名称”文本框中输入服务的名称，例如 webserver-svc，在“项目”下拉菜单中选择前面创建的 development 项目，单击“下一步”按钮进入“服务设置”对话框。

（2）在图 12-20 所示的对话框中，可以设置服务的访问类型、标签选择器以及端口等。

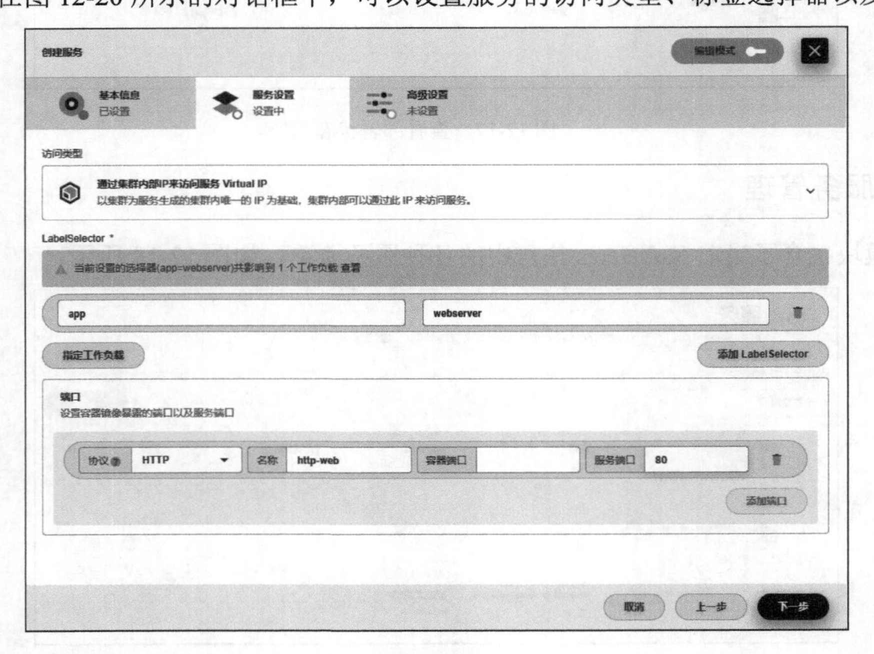

图 12-20 “服务设置”对话框

在本例中，服务类型选择“通过集群内部 IP 来访问服务 Virtual IP”选项。在 LabelSelector 的键文本框中输入 app，值文本框中输入 webserver，这个标签与前面创建的 Nginx 应用要一致。“服务端口”文本框中输入前面设置的 Nginx 对外服务的端口 80。最后点击“下一步”按钮。

（3）高级设置。在“高级设置”对话框中勾选“外网访问”复选框，然后在“访问方式”下拉菜单中选择 NodePort 选项，如图 12-21 所示。

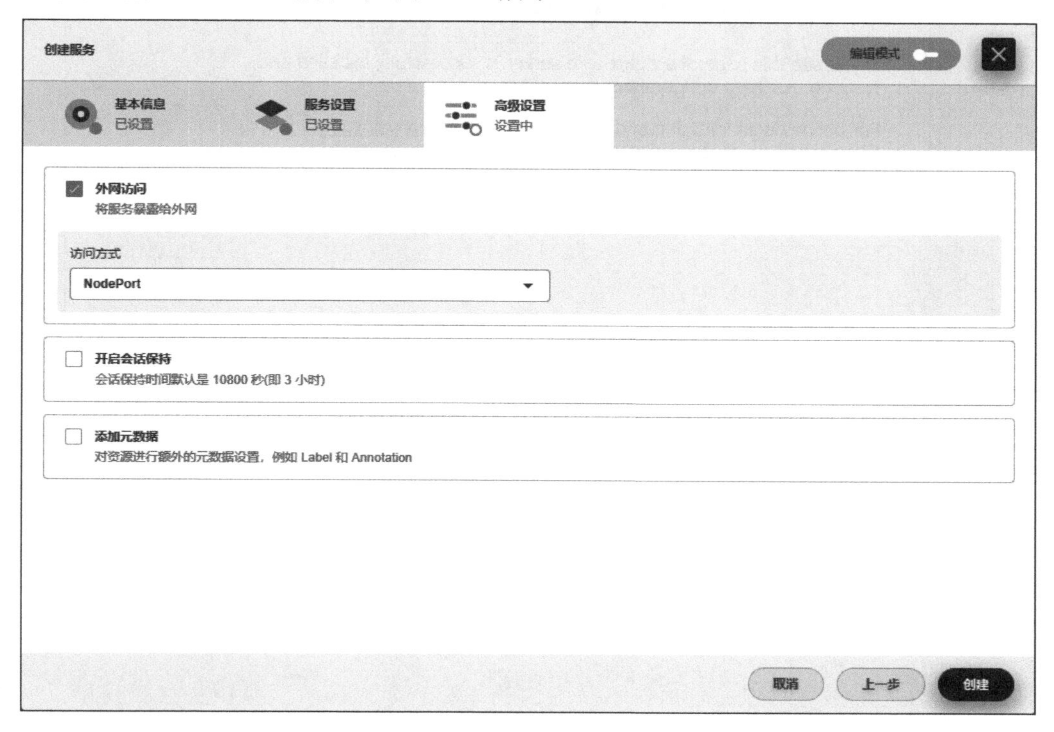

图 12-21　“高级设置”对话框

单击“创建”按钮，完成服务的创建。

此时，用户可以在服务列表中看到刚才创建的服务，如图 12-22 所示。

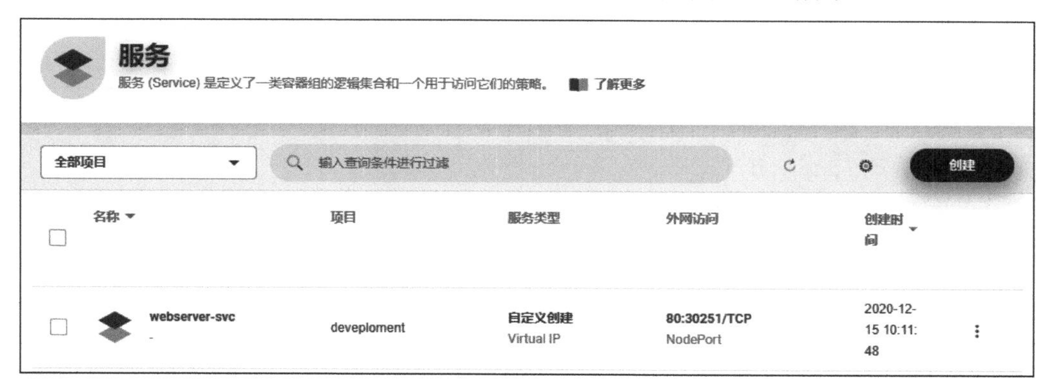

图 12-22　服务列表

从图 12-22 可以看到，Kubeletes 为当前的服务自动分配了一个端口 30251。用户可以通过节点的 IP 地址和端口 30251 来访问 Nginx 了，如图 12-23 所示。

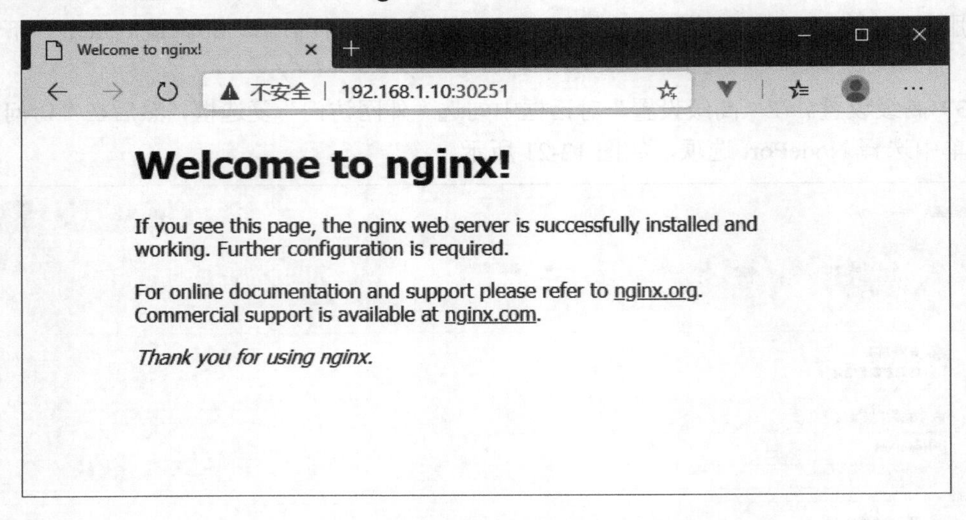

图 12-23　通过 NodePort 服务 Nginx 服务